M000110949

SCIENCE STUMPERS

Brain-Busting Scenarios....Solved with Science!

by
KEEGAN BURMARK
KEVIN J. BROUGHER

Missing Piece Press
TUCSON, ARIZONA

Other Publications from **Missing Piece Press**:

Thinklers!: A Collection of Brain Ticklers!

Thinklers! 2: More Brain Ticklers!

State Debate: 50 Unique Playing Cards and 50 Games for Learning About the States!

Number Wonders: A Collection of Amazing Number Facts!

Dreams, Screams, & JellyBeans!: Poems for All Ages

The Storybook: A novel for ages 10 on up.

Ways to Play with Words: A Collection of Word Games and Activities.

History Mysteries: A New Twist on Timelines!

Algebra Summary Sheets: Posters to Promote Proficiency

Frazzle: A Frenzied Game of Words

ShanJari: An African Game of Sequence and Strategy

Reindolphins : A Christmas Tale

Missing Piece Press is proud to have won the following awards:

- *Parents' Choice Gold*
- *Parents' Choice Approved*
- *Creative Child Magazine Seal of Excellence*
- *Teachers' Choice Award*
- *Family Choice Award*
- *Museum Tour Product of the Year Award*
- *Parent's Guide to Media Award*
- *GAMES 100 Award*
- *Tillywig Brain Child Award*

Science Stumpers: Brain-Busting Scenarios Solved with Science

Copyright © 2014, 2016 Keegan Burmark & Kevin Brougher

Printed in China

ISBN # 978-9703729-8-7

All rights reserved.
No part of this book may be reproduced by any means
without the written permission of the publisher.

Table of Contents

In every Stumper, interesting and important science concepts are presented and later explained. Readers cannot help but learn more about important scientific concepts. Though the Stumpers are not a cure-all for making science fun, they ARE a great way to get people thinking and learning in a fun way! Enjoy!!!

Branches of Sciences

This is not a complete list of the branches of science but includes many of the branches that the Stumpers book covers.

Aerodynamics: the study of the motion of gas on objects and the forces created

Anatomy: the study of the structure and organization of living things

Anthropology: the study of human cultures both past and present

Archaeology: the study of the material remains of cultures

Astronomy: the study of celestial objects in the universe

Astrophysics: the study of the physics of the universe

Bacteriology: the study of bacteria in relation to disease

Biochemistry: the study of the organic chemistry of compounds and processes occurring in organisms

Biophysics: the application of theories and methods of the physical sciences to questions of biology

Biology: the science that studies living organisms

Botany: the scientific study of plant life

Chemical Engineering: the application of science, mathematics, and economics to the process of converting raw materials or chemicals into more useful or valuable forms

Chemistry: the science of matter and its interactions with energy and itself

Climatology: the study of climates and investigations of its phenomena and causes

Computer Science: the systematic study of computing systems and computation

Ecology: the study of how organisms interact with each other and their environment

Electronics: science and technology of electronic phenomena

Engineering: the practical application of science to commerce or industry

Entomology: the study of insects

Environmental Science: the science of the interactions between the physical, chemical, and biological components of the environment

Forestry: the science of studying and managing forests and plantations, and related natural resources

Genetics: the science of genes, heredity, and the variation of organisms

Geology: the science of the Earth, its structure, and history

Marine Biology: the study of animal and plant life within saltwater ecosystems

Mathematics: a science dealing with the logic of quantity and shape and arrangement

Medicine: the science concerned with maintaining health and restoring it by treating disease

Meteorology: study of the atmosphere that focuses on weather processes and forecasting

Microbiology: the study of microorganisms, including viruses, prokaryotes and simple eukaryotes

Mineralogy: the study of the chemistry, crystal structure, and physical (including optical) properties of minerals

Molecular Biology: the study of biology at a molecular level

Nuclear Physics: the branch of physics concerned with the nucleus of the atom

Neurology: the branch of medicine dealing with the nervous system and its disorders

Oceanography: study of the earth's oceans and their interlinked ecosystems and chemical and physical processes

Organic Chemistry: the branch of chemistry dedicated to the study of the structures, synthesis, and reactions of carbon-containing compounds

Ornithology: the study of birds

Paleontology: the study of life-forms existing in former geological time periods

Petrology: the geological and chemical study of rocks

Physics: the study of the behavior and properties of matter

Physiology: the study of the mechanical, physical, and biochemical functions of living organisms

Radiology: the branch of medicine dealing with the applications of radiant energy, including x-rays and radioisotopes

Seismology: the study of earthquakes and the movement of waves through the Earth

Taxonomy: the science of classification of animals and plants

Thermodynamics: the physics of energy, heat, work, entropy and the spontaneity of processes

Zoology: the study of animals

Science Stumpers: Goals

Goal 1:   To raise awareness and increase knowledge of many important
          scientific facts and concepts.

Goal 2:   Accomplish goal number one in a FUN way!

This book may be used in many ways. We feel confident that however it is used, participants will be learning about scientific concepts. There has always been debate over what is important to know in regard to science. Students continue to ask, "Why do we need to know this?" It is true that the application of science is, at times, not as clear as the application of say, learning how to calculate compound interest.

Though, as E.D. Hirsch points out in his popular <u>Cultural Literacy</u> books, there are some things that people should know for no other reason than culture itself. A person can live a successful, enjoyable life never knowing much about how and why a light bulb works. However, somehow, someway, most Americans <u>*do*</u> know and most feel that all Americans <u>*should*</u> know.

This book will help ensure that scientific concepts will be learned. This book is not a complete science manual. Yet, it *will* help build a foundation for future learning. All the entries are great discussion starters or topics for further research. We hope that you enjoy using this book at home or at school.

And now...

The Stumpers!

Space Door

Bill and Maria enter a spaceship that will take them to the moon. Bill walks through a doorway that is just barely taller than him.   When in space, Bill unbuckles his seatbelt and floats to the door. He passes through the doorway and notices something strange.

Can you determine what he notices?

1. Bill can no longer fit in the doorway!   Without Earth's strong gravitational pull, his backbone expands. This makes him taller!

2. When standing upright on Earth, gravity is constantly compressing the twenty three discs that make up Bill's spine. However, in space these discs are allowed to expand. The lack of gravity provides 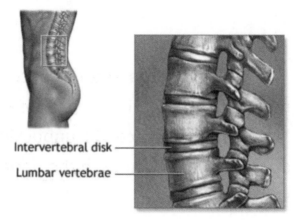 more room in between the vertebrae which water and other liquids can fill. This can result in a 3% gain in height that lasts for months even after living in normal gravity again! A similar process happens to people on Earth when they get a good night's sleep. The results are not quite as dramatic as the effects on astronauts but a sleeping person can be over an inch taller when they wake up after eight hours of sleep!

3. Many other physiological effects occur to astronauts who spend a significant time in space. Due to the lack of gravity, the body doesn't have to work as hard to keep many key structures of the body working properly. The body's bones decrease in mass because they are not being used to support the body's weight. Similarly, the body's muscles start to atrophy. The leg and back muscles which are used constantly in normal gravity for posture and keeping the body upright are not used at all and therefore decrease in size.

Intervertebral disk

Lumbar vertebrae

Moon Drop

Before he descended down the ladder to the surface of the moon, Bill decided to drop a few items down so he wouldn't have to carry them.   He had thrown most of what he needed over onto the soft surface of the moon. Lastly, he dropped a single piece of notebook paper (college ruled with three holes) and a writing pen. He released them from his hand at exactly the same time.

Can you determine which of these objects will hit the ground first?

1. Since there is no air on the moon, the pen and paper fall at exactly the same speed and hit the ground at exactly the same time!

2. Gravity pulls at the same rate on all objects if they are at the same location in the gravitational field. Even though the moon is not as large as the Earth, it does have mass so there is gravity. The moon's gravity will pull the two objects to the ground. However, without air resistance pushing up on both items, they will fall with the exact same speed.

   The resistance by air has nothing to do with the weight of the objects. To prove this, take two identical pieces of paper and drop them from the same height. Then, crumple one piece of paper into a ball and drop both again. Which one hit the ground first?

3. On earth, a pen and a piece of paper also have the same amount of gravity pulling down on each of them. However, if you drop both of these objects on Earth then the pen would hit the ground first. This happens because of air resistance. As the objects fall, the air molecules push up against the pen and paper and resist their downward motion. This slows down whichever object has the most surface area touching the air molecules. In this case the paper has more surface area so it falls more slowly.

Moon Scare

Maria and Bill are exploring the moon. Each has gone their own way. Maria decides to sneak around a crater and scare Bill by lighting off a string of firecrackers.   Maria pulls out her butane lighter and the firecrackers.

Can you determine the next series of events?

1. Nothing. Maria can't light the firecracker without air!

2. Earth's air is composed of 78% nitrogen, 21% oxygen, and 1% other gases. Maria's lighter needs the oxygen to create the chemical reaction that lights the fire!

3. The following is the chemical formula for the reaction that takes place with a butane lighter:

$$2\ C_4H_{10} +\ 13\ O_2\ \rightarrow\qquad 8\ CO_2\qquad + 10\ H_2O + Energy$$

Butane + Oxygen → Carbon Dioxide + Water + Fire

The Butane on the left side of the equation is useless without being combined with oxygen. When oxygen is available the reaction will occur and fire is created. The left over molecules of carbon dioxide and water float off into the air. Since the moon has no air, and therefore no oxygen, there cannot be any fire!

Moon Scare #2

Maria starts to think about scaring Bill again.   Fortunately, she is carrying two large cymbals with her.   She places one in each hand and slams them together.

If Maria is standing 10 feet away, can you determine or guess how long it will take for Bill to hear the clang and jump?

1. There is no air on the moon so there is no way for sound to travel from the cymbals to Bill's ears!

2. The force of the cymbals slamming together is normally transferred to air molecules. The air molecules bump against each other making a wave until their energy reaches Bill's ear. Bill's ear can then register the vibrations of the waves and he can hear the sound! On the moon there is no air to vibrate so Bill won't hear anything at all!

3. The waves in the air from the cymbals are called longitudinal waves. When a substance is moving back and forth in a horizontal direction it creates longitudinal waves (like an earthquake) while vertical waves are called transverse waves (like ocean waves). The longitudinal waves of sound hit the ear drum or 'tympanic membrane' and are then transmitted to the brain.

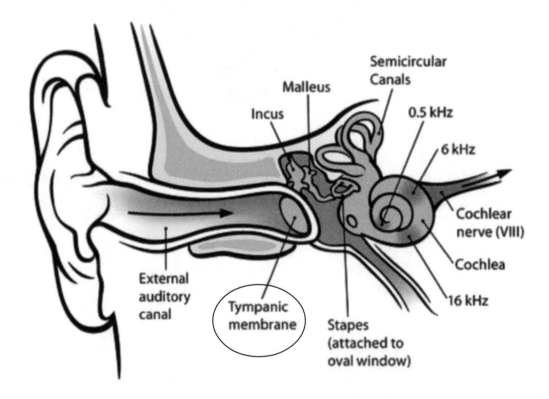

Moon Blast Off

Maria had felt silly that she had forgotten that there was no oxygen on the moon when she had tried to light the firecrackers. But, because there is no oxygen, she was starting to get worried that they would not be able to "light" the rocket engines to get off the moon!

Can you explain how Maria and Bill will get off the moon?

1. They carry their own oxygen with them!

2. There is no atmosphere containing oxygen in space. This means that the astronauts have to bring oxygen with them to cause a combustion reaction in the engines of the spacecraft. One way to carry oxygen is to convert it from a gas into a liquid. Liquid oxygen takes up much less space than gaseous oxygen which means that it can be carried easily on the spacecraft.

3. Liquid oxygen or 'LOX' is compact but it is also dangerously volatile. Because of this danger, astronauts like to carry their oxygen in the form of an 'oxidizer'. An oxidizer is a molecule that contains a high percentage of oxygen but is not dangerous to store. LOX is definitely the best oxidizer because it is composed of pure oxygen but the hydrogen peroxide that you most likely have in your bathroom cupboard is also an oxidizer! You can see the oxygen bubbles vaporizing from the liquid when you pour the hydrogen peroxide onto a wound. Hydrogen peroxide can also be used as an oxidizer in NASA rockets when LOX may be too dangerous!

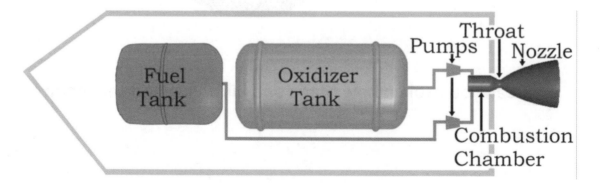

Birthday on the Moon

Maria and Bill's first day on the moon was Maria's birthday. Bill decided to surprise her with a few balloons. He filled the balloons up with air and gave them to Maria along with a nice card.

Can you determine or guess what happened when she took the balloons with her to the surface of the moon?

1. The balloons all explode as soon as Maria stepped onto the moon!

2. Without air pressure pushing down on the outside of the balloon the air on the inside of the balloon is free to push outwards as much as it pleases. This causes the balloon to rupture and explode!

3. There are two forces pushing in on the balloon; the elastic force of the balloon itself and the force of the atmosphere outside the balloon. The only force pushing out away from the balloon is the force of the air inside the balloon. Without the force of the air outside of the balloon the force of the air inside the balloon can overcome the elastic force of the balloon itself and it will burst!

$$F_{Elastic} + F_{Atmosphere} = F_{Air\ Inside\ Balloon}$$

The elastic force and the atmospheric force have to equal the force of the air inside the balloon in order for the balloon to be stable in equilibrium.

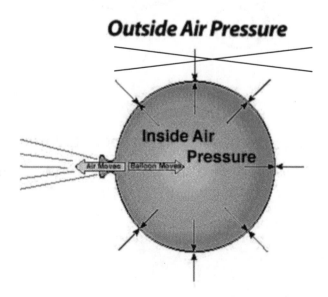

Birthday on the Moon #2

Bill apologized about forgetting his science and tried once again to give Maria some balloons. This time he took metallic balloons that do not stretch and a tank of HELIUM to the surface of the moon. He filled up the balloons and gave them to Maria, who was standing beside him.

Can you determine what happened when Maria let go of the balloons?

1. The balloons fell straight to the surface!

2. Helium balloons are buoyant in the atmosphere of the Earth. This means that the balloon and the air inside of the balloon are lighter than the air surrounding the balloon. Therefore the balloon floats in the air until something pulls it back down (like a string). However, the atmosphere of the moon is one hundred trillion times less dense than the atmosphere of the Earth. This means that the balloon on the moon will be much heavier than the air around it and will fall to the moon's surface.

3. The moon has very little atmosphere because its gravity is not strong enough to keep particles from escaping into space. The only reason the moon has any atmosphere at all is because of a constant flow of new particles hitting the surface from meteors and outgassing from beneath the moon's surface. These particles are quickly pushed out into space by the radiation of the sun and create a 'tail' of sodium and other particles behind the moon. This tail extends for millions of miles and is directed away from the sun.

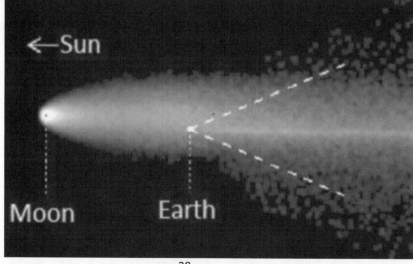

Moon Weight

Bill and Maria have both noticed that there is less gravity on the moon.   So, when they step on a scale, it shows that they weigh less then when they were back home on the larger earth.

If they traveled to another moon of the exact same size, could you determine, or guess, if they would weigh more, less, or the same as when they are on the earth's moon?

1. Even though it is the same size, it is all about the density of the moon. Density is how much "stuff" is packed into a given amount of space. If the moon was hollow, Bill and Maria would weigh a lot less. If the moon was filled with gold or lead, Bill and Maria would weigh a lot more!

2. Earth is the densest planet in the solar system with a density of 5.520 g/cm^3! The moon has a density of 3.346 g/cm^3. If the moon was the same size as it is now but had the <u>density</u> of the earth then Bill and Maria would feel almost twice as heavy!

3. Sir Isaac Newton figured out that the formula for gravity is as follows:

$$F = G \frac{m_1 \times m_2}{r^2}$$

This formula shows you that the force of gravity (F) between two object equals a gravitational constant (G) multiplied by both masses multiplies together (m$_1$ and m$_2$) divided by the distance between the object squared (r^2). You can see that the <u>size</u> of the two objects has nothing to do with the gravity! Only the mass and distance between them matters.

Vacation Speed

Bill and Maria had been planning a vacation for a long time. The day finally arrived for them to leave and they happily boarded a plane for the Caribbean Sea. It was a long flight and they all fell asleep. They woke up to find themselves traveling at over 1,000 miles per hour! However, they looked out the window of the plane and seemed to be at a complete stop.

Can you explain this situation?

1. They have landed but they have traveled to the equator and the earth is spinning (rotating) at roughly 1,070 miles per hour!

2. Speed of an object is completely relative to what is around that object. On earth we think that we are staying in the same spot when we are standing still but really we are moving at 1,070 mph. (at equator) We don't feel like we're moving that fast because everything around us is moving that fast as well. The same situation occurs in a car as you are traveling down the freeway. Every passenger in the car is traveling about sixty miles an hour but from your perspective the other passengers seem like they are not moving at all!

3. The reason that we do not feel like we are traveling at 478 meters/second (1070 mph) when we're standing still on Earth is due to acceleration. Scientifically speaking, acceleration is a change in velocity. As long as your velocity remains constant (0 mph or 1070 mph) you will have zero acceleration. The velocity that the earth spins is an impressive 478 m/s but that velocity is constant all the time so we feel nothing because there is no acceleration.

   Let's imagine for a second that the earth all of the sudden slows down to 1000 mph. Now the velocity has changed so you would feel the acceleration and you would be thrown across the ground at 70 mph!

Bike Ride Speed

It was a nice spring day. Bill decided to go for a bike ride.   He was on his bike for 1 hour.   When got home he calculated that he had traveled about 68,000 miles.

Can you determine what went wrong with Bill's calculations?

1. Bill took into account the velocity of the Earth in his calculations!

2. The earth is traveling around the sun at 67,062 mph. The earth is spinning at 1,070 mph. Jon rides his bike an average of 12 mph. If you add all these speeds together then you get 68,144mph!

       67,062 (velocity of the earth around the sun)

        1,070 (velocity of the earth spinning)

          12 (velocity of the bike on earth)

       68,144 mph (total velocity)

This means that if Bill rode his bike for one hour then his total distance travelled would be 68,144 miles! Not bad for one hours work!

3. The velocity of an object is relative to the other objects around it. We usually measure our velocity relative to the surface of the Earth but measuring velocity relative to the sun is just as valid. In fact, if Bill wanted to go even faster then he could measure his velocity in relation to the center of the Milky Way galaxy. With the galaxy as his reference point Bill would be going about half a million miles per hour.

Space Plane

Bill and Maria helped design a space plane that can travel at the speed of light. A test pilot is chosen for the first test flight.   The pilot takes off from the moon and in a split second is traveling like a light beam at 186,000 miles per SECOND!   The auto-pilot brings the craft back to the moon.

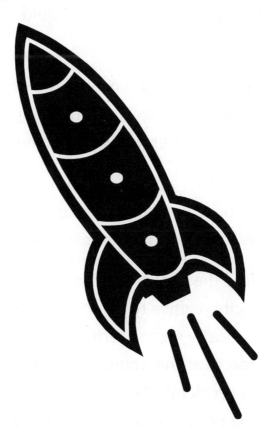

Can you determine what the pilot might have
to say about this monumental flight?

1. Sadly, the pilot is unable to say anything about the trip. The incredible acceleration killed him instantly.

2. When the spacecraft leapt to light speed, the pilot would have been thrown against the wall of the spacecraft with tremendous force. Sir Isaac Newton discovered that we measure force using the equation force equals mass multiplied by acceleration (F=ma). If the pilot's mass is 73 kilograms (about 160 lbs) and his acceleration is 300,000,000 m/s^2 (the speed of light) then the force on his body would equal 21,900,000,000 Newtons of force! To compare, a punch in the face would be about 45 Newtons of force.

3. Sir Isaac Newton also discovered the property of inertia. Inertia is the resistance an object has to changing it's state of motion. Newton put it a different way when he wrote his first law of motion:

   "The velocity of a body remains constant unless the body is acted upon by an external force."

   Right before the pilot starts the engine he is traveling with the same velocity relative to his spacecraft. Remember that constant velocity means zero acceleration so the pilot doesn't feel any forces at this point. After the engine is turned on the spacecraft accelerated rapidly but the pilot's body wants to stay in the same place due to inertia. His body pushes back on the spacecraft just like your body would push up against the seat of a car as you accelerate onto the freeway. The only difference is that the plane is going much, much faster and therefore has much, much more force!

Space Plane # 2

Bill and Maria want to continue experimenting with their light-speed space plane.   They make adjustments so the plane GRADUALLY accelerates to the speed of light.

This time, a new test pilot takes off from San Francisco. The plane accelerates quickly and the pilot feels the increased 'G' force but it is bearable. The pilot's plan is to reach an altitude of 30,000 feet, fly for 30 minutes, then return to San Francisco.

Can you determine, or guess, how many times
his plane will circle the earth?

1. None! The plane would burn up almost instantly.

2. As Bill and Maria's light speed plane climbed in altitude it would start to get colder due to the colder ambient temperature in the atmosphere. However, as the plane increased in speed this cooling effect would matter less and less. Initially the ambient temperature will be the main determinant of the temperature of the plane but as the plane picks up speed the friction and pressure of the air will start to play a bigger and bigger role. When the plane hits the air molecules the kinetic movement energy is converted into heat energy. This causes the plane to heat up more and more as its speed increases. Eventually the temperature of the plane will get so high that it will melt the hull!

3. There are many variables when trying to calculate how hot a plane will get at high speeds including air density, elevation, specific heat, and the duration of the flight. However, there is a mathematical formula to help make estimates of the temperature at different speeds:

$$T_0 = T + \frac{V^2}{2C_p}$$

In this formula T_0 is the temperature of the plane's hull, T is the ambient air temperature, V is the velocity of the plane, and C_p is the specific heat of the plane. This means that an SR-71, one of the fastest planes ever built, travelling at Mach 3.5 would have a hull temperature of almost 900°F! Mach 3.5 is only 2,664 mph. Light speed is 670,616,629 mph! No known substance could come close to withstanding light speed temperatures.

Space Plane # 3

Bill and Maria decide to give it one last try. They recruit one last test pilot. The pilot takes off from earth, reaches space then slowly accelerates to almost the speed of light. The pilot looks at her watch and after an hour she begins the slow down process. (She learned that she can't DECELERATE to quickly either!)   Finally, she lands back on earth.

Can you guess what might be the first thing she
shares with Bill and Maria about the trip?

1. Bill and Maria have died of old age by the time she gets back!

2. As the pilot's speed increases, her TIME is also slowing down. Even though the pilot feels like she has only been on the ship for an hour, her friends back down on earth see her orbiting the earth for decades. Einstein predicted this when he thought of his theory of relativity. The passing of time is **relative** to how fast you are traveling in space compared to somebody else.

3. To find exactly how much time passed for Bill and Maria relative to the pilot you can use the 'Lorentz Factor'. The graph below shows you how the Lorentz Factor relates the pilot's velocity to her passage of time:

   As the pilot's velocity increases toward the speed of light (x-axis), her speed of time (y-axis) relative to earth is increasing at an exponential rate. This means that for objects moving very slowly, like Bill and Maria on earth, time will seem to be the same for both objects. However, the closer you get to speed of light the faster time accelerates and the more dramatic the time difference will be!

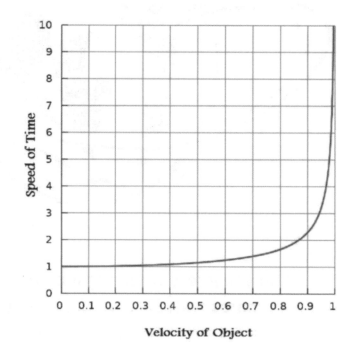

Plane Weight Problem

Bill has loved flying since he was kid. Today he gets to pilot the fastest plane ever built – the SR-71 Blackbird!   The pilot seat is equipped with a brand new HIGHLY sensitive and accurate scale.   He buckles in and checks his weight, then takes off into the clear sky.

When he has reached top speed, he checks his weight again.

Can you determine or guess if his weight has changed?   If so, why?

1. When flying in his Blackbird, Bill will seem a tiny bit heavier than if he were simply standing on the ground.

2. Gravity from the earth pulling on the mass of your body is what causes your weight. This is why your weight can change depending on where you are. For example, in space you are almost weightless! However, your mass doesn't change when you are traveling at a higher speed. What does change is the space around you which makes you seem heavier than you are. This is why Bill will seem heavier at a higher speed. If he increases his speed to near the speed of light then he would begin to get infinitely heavier!

3. Physicists are still arguing about the definition of the word 'mass'. Some physicists, like Einstein, think that mass should be called 'relativistic mass'. Believers in this theory about mass think that mass increases with increasing velocity. This is compatible with many of Einstein's formulas. However, some other physicists believe that mass should be called 'invariant mass'. Believers in the theory of invariant mass think that mass does NOT change with increasing velocity. This type of mass is more compatible with many mathematical equations.

Rocket Fuel

Bill and Maria's engineers have built a rocket engine that uses ANY matter as fuel. Rocks, trees, water, dirt, old cars, old planes, old trains... ANYTHING can be used as fuel to make the rocket go faster.

Can you determine how many pounds of fuel it would take to have this rocket reach the speed of light?

1. There is not enough matter in the whole universe!

2. The graph of the energy it takes to move an object compared to that object's speed is shown in the figure below. You can see that the line next to 'Relativistic Mass' is a logarithmically shaped graph. Logarithmic functions increase faster and faster instead of in a straight line. In this case, as the velocity of the rocket gets closer and closer to the speed of light (c) the amount of energy it takes to push the rocket increases immensely!

3. The reason that there isn't enough energy in the universe to get the rocket to light speed is because the theory of relativistic mass says that objects get heavier the faster they move. The upper line of the graph below shows us that if this is true then when we get very near to light speed (c) then the amount of energy we need to keep pushing the rocket will be tremendous!

If the theory of invariant mass is true then the lower line of the graph would be more accurate. This line shows us that it could be possible to reach light speed as long as mass stays the same at all velocities.

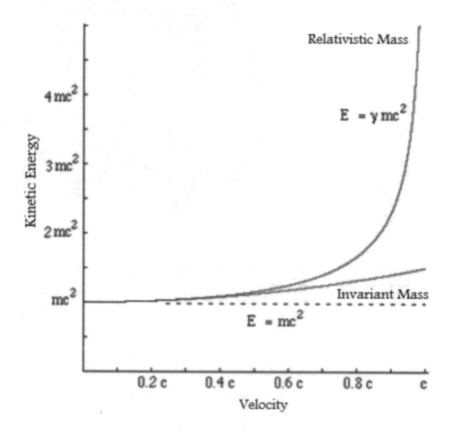

No Spin Planet

The earth has been spinning around its axis since it was formed.   It has been gradually slowing down ever since.

Can you determine what would happen to a farmer Joe's apple orchard if the earth were to **<u>suddenly</u>** stop spinning?

1. They would all topple down to the ground!

2. The Earth is spinning at 1667 km/hr (1036 mi/hr) at the equator. If the Earth itself were to stop spinning all at once then everything on top of the Earth would still be travelling at 1667 km/hr! Newton's first law says that an object in motion will stay in motion unless another force acts upon it. So, all the trees (and humans) on Earth would continue travelling in the same direction if the Earth stopped!

3. Momentum is the quantity of movement that an object has based off of its mass and its velocity. Because velocity is a vector and has a direction, momentum is also a vector in the same direction as the velocity. The formula for momentum is as follows:

P (momentum) = m (mass) X v (velocity)

If the world were to stop turning then everything on earth that was previously still (p=(mass)(0)) would gain momentum proportional to their mass and 1667 km/hr (p=(mass)(1667 km/hr))! This would feel similar to slamming on the brakes of a car going 1667 km/hr without wearing a seatbelt!

Another strange effect of the Earth ceasing to spin would be that the earth would become perfectly spherical since there would be no centrifugal force bulging out the equatorial areas like is currently happening.   This change in mass would result in a change in gravity which would cause all the planet's oceans to congregate around the equator. This would create one, very large, equatorial ocean!

No Spin Planet #2

Miraculously, farmer Joe survived the quick stop of the earth. Joe picked up the apples that had spilled out of crate.   He put the crate back on to his very accurate scale.

Joe had the exact same apples in the crate as when the earth was spinning.   Nothing else had changed.

Can you determine what Joe noticed and why?

1. The apples weighed slightly more than they did when the earth was spinning!

2. Since the earth is spinning, everything on its surface has 'centrifugal force'. This force lifts everything on earth's surface up with just a few Newtons of force. Farmer Joe doesn't even notice this lifting because it is constant all the time. One example of centrifugal force is standing on a merry-go-round. On the merry-go-round you feel a force pulling you to the outside. The faster the merry-go-round goes, the more centrifugal force is pulling you. Eventually it would pull with more force than is keeping you on the ground and you would go flying off!

3. Since the centrifugal force of our spinning earth lifts us up just a tiny bit we are pushing down on the earth is less force. But exactly how much force is pulling on us? The amount of force varies by the mass of the object so let's use farmer Joe as our example:

   The formula for centrifugal force is:

   F_c = centrifugal force on Joe

   $$F_c = \frac{mv^2}{r}$$

   M = Joe's mass (≈70 kg)

   v^2 = Velocity of Earth's rotation squared (≈465 m/s)^2

   r = Radius of the Earth (≈6.37x10^6 m)

   When you plug these numbers into the equation above you should find that Farmer Joe is normally being lifted up with an extra 2.376 Newtons when the earth is spinning.

Old Age

Kevin & Keegan are born on the same day. They both live a productive life and then die on the SAME day too.   Yet, Kevin dies when he is 90 years old and Keegan dies when he is 3 years old.

Can you explain their difference in age?

1. Keegan lived on Saturn while Kevin lived on the Earth!

2. A year on Saturn is equal to about thirty earth years. So three years on Saturn is the same thing as ninety years on Earth. This doesn't mean that Kevin lived longer than Keegan. They both lived the same amount of time but Saturn is much further away from the sun so it takes much longer than Earth to make one full rotation. The Earth is also moving at a much faster rate than Saturn. Earth has an orbital speed around the sun of 29.8 km/sec while Saturn has an orbital speed of only 9.7 km/sec.

3. Planets that are closer to the sun have higher orbital speeds. This was most likely caused by the centripetal force of the beginning of our solar system. When dust and rocks were condensing into bigger bodies like planets they tended to go faster the closer they were to the sun's strong gravitational pull. This is similar to a figure skater doing a twirl. The skater starts by spinning slowly with their arms or legs far from their body. As they pull their arms and legs in closer, their spinning get

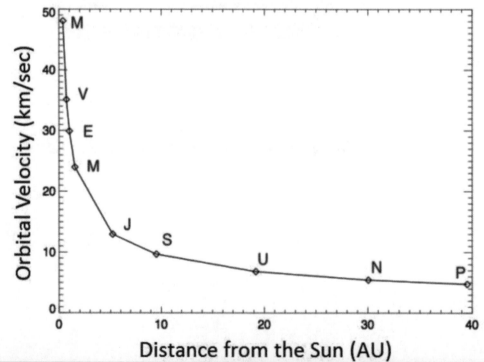

Meteors

Scientists created a large net to catch incoming "falling stars", better known in the scientific community as meteors.   They took the first five meteorites that they caught back to a storage building.   All of the meteorites were of average size and weight.

Can you make a close estimate to how much area they would need to store these five meteorites?

1. The estimated size of an average meteorite hitting Earth is .2 mm. So five meteorites would need about 1 mm of space. This is about the size of the point of your pencil!

2. When we see a space rock coming into our atmosphere it is called a meteor. Meteors shine brightly because the rocks that meteors are made out of are actually burning up as they enter our atmosphere. The friction of our atmosphere hitting the space rock causes it to be superheated into what we can see as the meteor. When meteors actually hit the Earth's surface they are much smaller because most of their mass has burned up in the atmosphere. Once they hit the ground we call them 'meteorites'.

3. The amount of mass lost by a meteor as it travels through Earth's atmosphere depends on two factors. The first factor is the angle that the meteor travels as it travels through the atmosphere. The lower the angle of the meteor, the longer the meteor has to travel through the atmosphere and therefore more mass will be burned off of the meteor.

The second factor is the size of the meteor. Bigger meteors will create bigger meteorites. The graph on the right shows how often different sized meteorites hit the Earth.

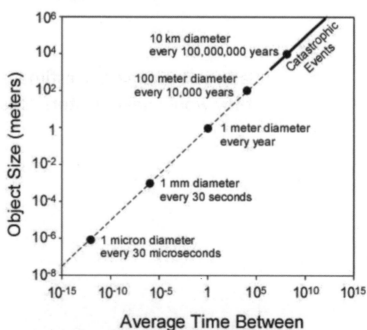

RC Plane

Maria brought a remote controlled plane with her to the moon. She cleared a 20 foot long strip for a runway.

She started the battery operated engine (no oxygen for a gas engine) then gave it full power. The propeller was spinning at full speed and the brakes were off yet the plane was not moving.

Can you determine Maria's problem?

1. There is no air for the propeller to push to give the plane the propulsion it needs to move!

2. A propeller is able to push the plane forward by pulling the air in front of the plane towards the back of the plane. This creates a force called thrust which propels the plane forward. The thrust of the propeller does not actually lift the plane into the air. The wings of the plane are what accomplish that feat. Without air the propeller has nothing to pull and therefore there is no thrust!

3. The speed of a propeller engine plane is controlled by the propellers RPMs, revolutions per minute. The faster the propeller rotates the faster the plane's velocity will be. However, propeller propelled planes can only go around 540 mph at their absolute maximum velocity. Around the time of World War II many planes switched over to jet propulsion systems. Jet propulsion systems, like the ones powering a Boeing 747, also pull air from the front of the plane to the back to create thrust. However, in a jet engine the air entering the engine is compressed and then heated up to extreme temperatures. The difference in temperature causes the air to shoot out the back side of the engine at a very high velocity. This difference in velocity of air coming into the engine compared to the air exiting the engine causes much more thrust than propeller propelled engine.

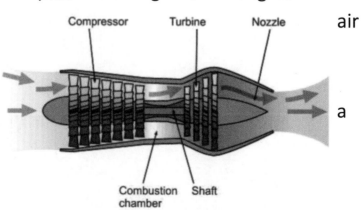

Compressor   Turbine   Nozzle

Combustion   Shaft
chamber

RC Plane #2

Maria made changes to her plane. She replaced the propellers with rocket engines.   The plane quickly picked up speed and zoomed down the runway. Maria could see the flaps move as she used the remote control but no matter what she did the plane did not take flight.

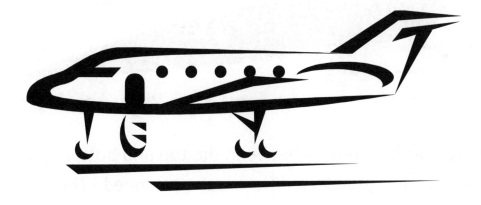

Can you determine what the problem is now?

1. Even though the rocket engines are pushing Maria's plane forward, there is no air to travel over the wings and create lift!

2. In all fixed winged planes, lift is achieved by creating an air pressure difference between the top and bottom of the wing. The wing's airfoil design causes high pressure below the wing and low pressure above the wing. The wing, and the plane itself, wants to move towards the low pressure and away from the high pressure thereby creating lift!

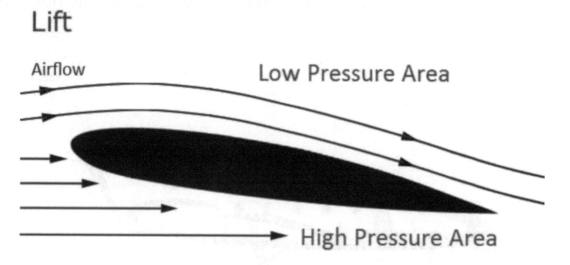

3. The pressure differences between the top and bottom parts of the wings are caused by a difference in air speed. The air speed over the top of the wing is much quicker because the air has a longer distance to travel than the air below the wing.

The figure on the right shows how the air separates as it flows over the top and bottom of the wing.

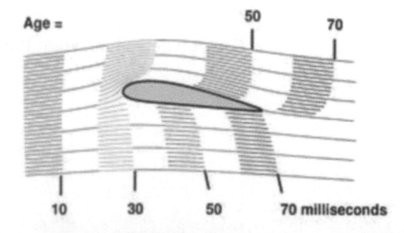

Fire Proof

Scientists have created a spaceship and spacesuits that can withstand the heat of the sun and adjust the gravity to make it the same as on earth.

The 93 million mile trip takes a LONG time but, they finally arrive. The astronauts prepare for landing.   Upon touchdown they realize that they had forgotten one important fact.

Can you determine what this might be?

1. The surface of the sun is not a solid so the spacecraft sinks right into the sun!

2. The sun is not made of a solid, liquid, or gas. Instead it is made of a state of matter called 'plasma'. Plasma is created when a substance is heated to extremely hot temperatures like on the surface of the sun where the temperature is almost 10,000° F. This high temperature breaks all bonds between atoms. It also causes electrons to be extremely energized and leave their atoms! Plasmas are rare on earth but they can be seen in objects like neon signs and in natural phenomena like lightning strikes.

3. Plasma occurs when matter, usually a gas, is ionized into its nuclei and electrons. Electrons are negative and nuclei are positive but the plasma as a whole is usually uncharged. Each substance has a different temperature that it must reach in order to turn into a plasma. This is similar in concept to the melting or boiling point of a substance but the temperature needed to make a plasma is much higher.

Plasmas make up the majority of mass in or universe. Stars and interstellar nebulas are usually 99.99% composed of plasmas. Our solid, liquid, and gas states of matter are actually quite rare!

On Neptune

Bill and his family have just finished breakfast. They decide to take a morning dip in the backyard pool. After a few hours Bill does one last cannonball off the diving board then tells the family that he has to go into work.

Bill arrives at the World Space Agency and walks through a portal that instantaneously transports him to a housing unit that the agency has built on Neptune.   Bill uses the super-high-powered telescope to look back at Earth. He zooms on his house (it's a GOOD telescope).

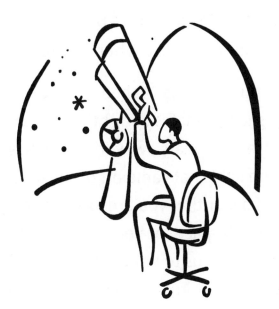

Can you determine what he might see?

1. Bill would see himself playing in the pool four hours earlier because the light from earth takes four hours to reach Neptune!

2. Light travels at about 300,000 km/sec. That is the equivalent of going around the earth seven and half times in one second! The distance from the Earth to Neptune is an expansive 4,347,000,000 km. This means that the sunlight bouncing off of Bill as he plays in the pool will take four hours to travel those four billion kilometers before reaching the eyes of Jason on Neptune.

3. Humans are primarily visual creatures. This trait comes from our ancestors who lived on the Savannahs of Africa and had to see camouflaged predators coming from across the vast landscape. Because we are so visual we rely highly on our sight to interpret the world around us. We think of the things that we see as absolute reality when really we are seeing the past.

   Think about this example. When you look up at the stars at night you are not seeing what is currently happening in the universe. Most stars are light-years away from Earth, which means that you are seeing the light that came off of those stars years ago. The light coming off of a star that you are looking at right now has been traveling through space for years before it hits your eye. This is a strange concept for us to understand because when we look at objects like this book in everyday life the light is so fast that we don't notice any delay at all. In reality you are not seeing this book page right now. You are actually looking at the light that bounced off this book page about .00000000167 (1.67×10^{-9}) seconds ago. Everything you see is from the past!

The Race

Cliff was born with no legs. Yet, on a recent race, he was clocked at almost 12 miles per hour. He was not in a wheel chair and was not aided by anyone or anything else. He did not even use arms!

Can you determine how he achieved this speed?

1. Cliff is a Black Mambo snake!

2. Many people think that snakes move by pushing off of objects with their coils. This is actually not completely true. Snakes can move forward with no arms or legs because of their special scales. Snake scales seem very smooth when you feel them. However, if you were to rub a snake in the opposite direction of its scales then you would find the snake's scales are very rough. This lets them glide forward easily and also get traction against the ground when they move side to side. This type of movement is called "Serpentine."

3. There are three other types of snake movement shown in the diagram below. Each type of movement has been evolved by different snakes to best fit their type of environment. For example, 'Sidewinding' is the best strategy for desert snakes because it is an efficient way of moving through crumbly, dry sand environments.

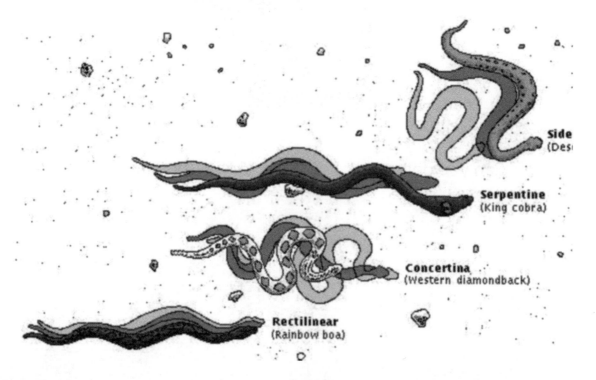

Side
(Des

Serpentine
(King cobra)

Concertina
(Western diamondback)

Rectilinear
(Rainbow boa)

Burning Question

Bill and Maria have a 10ft by 10ft by 10 ft glass cube that weighs one hundred pounds. The cube is airtight and is sitting on a large, very accurate scale.

Inside the cube there is an electronic igniter and a small pile of wood. The wood weighs exactly five pounds - the igniter weighs one pound. The total weight of the cube with the wood and igniter inside is 106 pounds.

The igniter lights wood on fire in the cube. There is enough oxygen in the cube for all of the wood to burn.

Can you now determine the approximate weight of the cube?

1. The cube still weighs 106 pounds!

2. When the wood is burned in the cube it seems to turn into a pile of ash that is much smaller than the original wood. This is the same thing that would happen to the wood in a campfire. However, the Law of Conservation of Mass tells us that all of the mass we started with still exists after we burn the wood; it is now just in a different form. When wood is burned it turns into ash and Carbon Dioxide. The Carbon Dioxide (CO_2) floats into the air. Since the air in the cube is contained there is no mass lost and therefore the weight of the cube will be the same before and after the burning of the wood!

3. Wood is made up of a sugar called 'Cellulose' which is made up of many Glucose monomers linked together. If we take one sugar molecule to represent our wood then our chemical formula would look like this:

$$C_6H_{12}O_6 \quad + \quad 6\,O_2 \quad = \quad 6\,CO_2 \quad + \quad 6\,H_2O$$

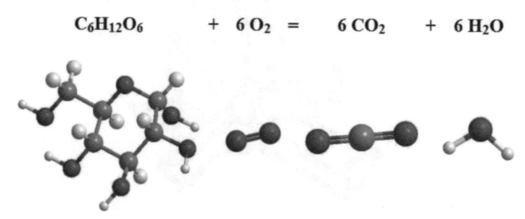

Wood and **oxygen** on the left combine to make **carbon dioxide** and **water** on the right. If you add up all of the different atoms on each side you will see that both sides are equal. If no mass is lost from burning the wood then the weight will stay the same even though the wood changed from a solid state to a gaseous state.

Take Off

Bill is in a plane in Phoenix, Arizona.   The weather is clear skies and 100°.

Maria is in a plane in Anchorage, Alaska. The weather there is clear skies and 35°.

The plane, plane weight, runway, and humidity are ALL exactly alike.

Can you now determine which plane would take the LEAST amount of runway to get off the ground?

1. Maria's plane would need the least amount of runway because colder air is denser than warmer air!

2. One factor that controls the lift of the plane is how much air is traveling over and under the wing. If there are less air molecules then there is less mass to create force to lift the plane. When the weather is very warm there is more Kinetic energy exciting the air molecules to move faster. When air moves faster it hits other air molecules at these high speeds. Because the speed of the air molecules colliding has increased, it causes the air molecules to move away from each other. This means that warm air is less dense than cool air and will therefore take Bill longer to take off than Maria!

3. So Maria is able to take off in a shorter amount of time, on a shorter span of runway, because Alaskan air is denser than Arizonan air. The formula for density is as follows:

$$density = \frac{mass}{volume}$$

The density of the air is determined by the mass of the air (the number of air molecules) divided by the volume that the air takes up. In this situation the mass will stay the same. The number of molecules in the air does not change. However, the atmosphere is a very large place and can expand as much as it wants into space. Therefore, in a hot place the mass will stay the same but the volume will increase. When we increase the denominator of a fraction the density will get smaller. Less density of air means less air going over the wings, and less lift.

High Heat Take Off

An aircraft carrier is cruising the waters of the Indian Ocean. The air temperature is hovering around 100 degrees.

On the flight deck there are life boats, safety nets, fire hoses, and tool boxes. It has been calculated that with the high temperature, there is not enough runway to take off.

Yet, one sailor devises a way to get the jet in the air.

Can you now figure out his plan?

1. Use fire hoses to shoot water through the jet engines!

2. The solution to this scenario is similar to the last but uses a different state of matter. The amount of lift that a jet gets depends on the amount of matter that it can pull through its jet engines. Usually this matter is air but if we add some water then we will be increasing the amount of mass the engine can pull in. If we increase the mass being pulled in by the engine then we have increased the amount of force, or thrust, exerted on the plane. This gives the plane more power to move forward and lets the plane take off with a smaller amount of runway!

3. There is a limit to how much water a jet engine can take in to increase thrust. Refer to the diagram of a jet engine on page 54. If there is too much water entering the engine then the combustion chamber doesn't get enough oxygen and the combustion reaction can't occur. Another reason that you can't have too much water in the engine is that the water floods the combustion chamber and results in the igniter being drowned. This is termed a 'flame out' because the combustion has ceased. All jet engines are tested to withstand a high amount of water intake because most jets at some point fly through clouds. Most engines can withstand as much as 45% water intake as long as their engine is operating with enough power! Water injection take-offs cause a lot of smoke, as you can see here.

Atom Counting

Bill didn't have much to do so he used an electron microscope and started counting the number of atoms in a typical – very average size – single grain of sand.

If he counted one atom per second, can you give a reasonable estimate on how long it would take him to count every atom in grain of sand?

1. 2,378,234,400,000 years! That is over 2 trillion years!

2. In order to find how many atoms there are in sand we first have to see what <u>types</u> of atoms are in sand. All atoms are not created equal. Some are bigger and some are smaller. Sand is mostly composed of Silica which has the molecular formula SiO_2. This means there are two Oxygen atoms and one atom of Silicon in each molecule of sand. By seeing how many times this molecule fits into one grain of sand we can find our number of molecules.

3. To find how many molecules of Silica there are in a grain of sand we will need to find the weight of Silica. Silicon weighs about 28 atomic units (AUs) while Oxygen weighs about 16 AUs. This means Silica weighs 60 AUs total.

   Next we'll need to use Avagadro's Number which is 6.023×10^{23} AU's per gram.

   $$\frac{6.023 \times 10^{23} \text{ AUs}}{1 \text{ gram}} \times \frac{1 \text{ Molecules Silica}}{60 \text{ AUs}} = \frac{1 \times 10^{22} \text{ Molecules Silica}}{\text{gram}}$$

   Now we know the amount of molecules of Silica per gram of sand. If we have one mm^3 of sand it will weigh about .0025 grams. We can use this information to find the problem:

   $$\frac{1 \times 10^{22} \text{ Molecules Silica}}{\text{gram of sand}} \times \frac{.0025 \text{ grams of sand}}{1 \text{ grain of sand}} \times \frac{3 \text{ atoms}}{1 \text{ Molecule Silica}}$$

   When we multiply this together and cancel out the 'molecules of silica' and 'grams of sand' from the top and bottom of the equation we find that there are 7.5×10^{19} atoms per grain of sand. If it takes one second to count each 7.5×10^{19} atoms then you get 2.38×10^{12} years to count all the atoms in one grain of sand!

Oxygen Level

The air that we breathe is about 21% oxygen.

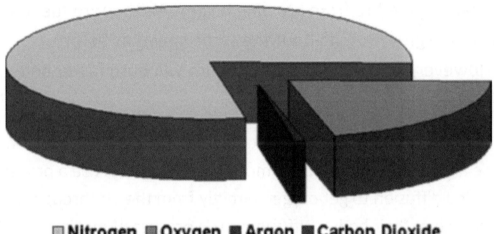

☐ Nitrogen ☐ Oxygen ■ Argon ■ Carbon Dioxide

Can you determine what significant change would occur if the percent of oxygen was increased to 30%?

1. Most of the Earth would burst into flames!

2. With more Oxygen in the atmosphere there is an increased risk of objects catching on fire. Fires physically need two components to burn properly; fuel and Oxygen. Usually there is enough fuel for a fire to burn because most objects on Earth are flammable. Therefore, Oxygen is usually the limiting factor to the growth of a fire. A limiting factor is something that is stopping the fire from growing bigger. With more fuel the fire will still grow but at about the same speed as before. However, with more Oxygen the fire will burn faster and hotter!

3. Another consequence to an increase in atmospheric oxygen would be an increase in the size of most insects. Insects use a process called diffusion to get oxygen directly from the air through the surface of their body. Diffusion is the process of moving a substance from a higher concentration to a lower concentration without using energy. This process works well for insects because they don't have to build and operate energy expensive lungs. However, the drawback to diffusion is that it only works over a relatively short distance. If atmospheric oxygen levels were higher then diffusion of Oxygen into insect cells would happen more quickly and travel further into their bodies. This would mean that an insect could grow bigger and still get the necessary Oxygen to the cells in the middle of its body.

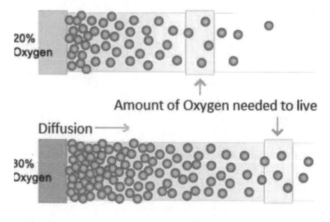

Window Wonder

A large thick piece of glass was constructed as a window for an industrial building. It measured exactly 10 feet long by 10 feet wide by 1 inch thick.

After 100 years of use, scientist went back to measure the window.

Can you determine if the dimensions had changed and why?

1. The glass has stayed exactly the same width throughout the entire window!

2. Many people today believe that glass is actually a very slow moving liquid but this is not true. Most of these beliefs trace back to a story from Europe where scientists examined glass from cathedrals that were hundreds of years old and found that the glass at the bottom of the cathedral windows was thicker than the glass at the top. It turns out that this thickness at the bottom of the window is actually the result of imperfections in the ancient techniques used to make the glass windows and is <u>not</u> caused by the glass itself moving.

3. The reason that this urban legend has persisted for so long is because glass actually does have the properties of a liquid and of a solid. Most solids have a crystalline-like structure that is very ordered and symmetrical. However, glass has a structure that is more random and not symmetrical at all. Scientists call this an 'amorphous solid'. The "**a**" means "*NO,*" "**morph**" means "*SHAPE,*" and "**ous,**" means "*PROPERTY OF.*" Therefore an amorphous solid is a solid that does not have a definite shape.

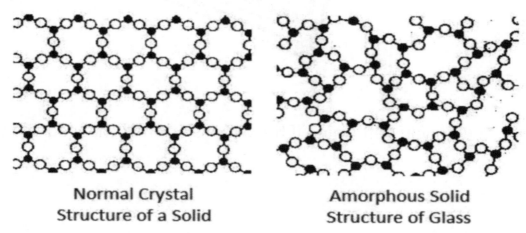

Normal Crystal
Structure of a Solid

Amorphous Solid
Structure of Glass

Space Weights

Bill and Maria enter the new space station orbiting hundreds of miles above the earth.

They notice that in one part of the space station they can easily lift and move a 200 pound container. Yet in another part of the space station they cannot. The container is not attached to the floor in any way and there is no magnetic force being used.

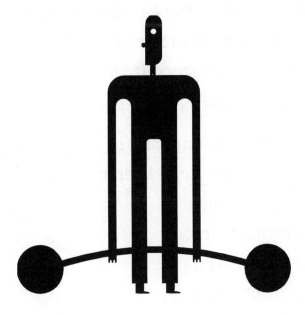

Can you determine what might be the cause?

1. Part of the space station is spinning and creating artificial gravity!

2. Gravity is usually a very small force in open space which is why astronauts feel weightless during space travel. However, weightlessness for long periods of time can cause many health problems like muscle loss and deterioration of bones. To solve this problem artificial gravity could be created. If a space station is spinning then there is a centrifugal force pushing everything outwards. An example of a centrifugal force would be holding a friend's hands and spinning in a circle. If you were to let go of your partner then you would both go flying away from each other. This is how artificial gravity is created. The spinning of the spacecraft forces everything in the spacecraft to the outer part of the spacecraft. In effect, it would be like standing on the ceiling!

3. Centrifugal force is the opposite of centripetal force. Centripetal force is the force keeping something moving towards the center when spinning. Centrifugal force is an inertial force that seems to push outwards on a spinning body like that of an astronaut. The hull of the space ship pushes back against the astronaut creating the illusion of gravity.

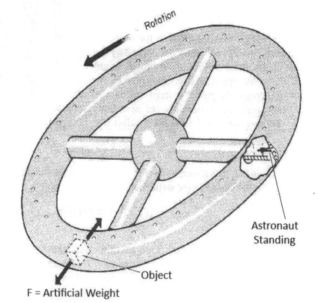

Rotation

Astronaut Standing

Object

F = Artificial Weight

Furnace Choice

Bill was the lucky winner of a carnival raffle. The winning prize was either a nuclear powered furnace with 1 pound of nuclear fuel or a coal burning furnace with 16,000 pounds of coal.

If Bill wanted to choose the furnace that would warm his house the longest before he had to buy more fuel, which one should he choose?

1. Both of those furnaces contain the same amount of energy!

2. Even though the amount of energy between the two furnaces is equivalent, Bill would be smart to take the nuclear powered furnace. Bill would need a large semi truck filled up completely to transport 16,000 pounds of coal! However, only one pound of nuclear fuel can fuel a large icebreaking ship that has a 70,000 horsepower capacity for an entire day. Compare that to your car or school bus!

3. Burning coal releases the chemical energy that is stored in the bonds of the hydrocarbon coal molecules. Storing energy in chemical bonds is a common and useful way to store energy for transportation. However, nuclear energy doesn't just break bonds, it converts matter directly into energy. Einstein's most famous equation explains why this is true:

$$E = mc^2$$

Einstein said that energy (E) is equal to the mass of an object (m) multiplied by the speed of light (c) squared. This equation shows that a little bit of mass can be converted into a LOT of energy. Normally this energy is locked up in the matter of objects and is relatively stable. Very small amounts of nuclear energy can be released naturally when cosmic rays hit the earth. When we use nuclear power plants and bombs we are artificially creating a nuclear reaction and causing some of the mass of plutonium or uranium to turn into energy.

Air Quality

Maria walks into a sealed room containing 78% nitrogen.

Can you explain what effects Maria would experience and the time that these effects would begin and end?

1. Maria would have no negative effects because our air is already 78% Nitrogen!

2. Nitrogen is an element that all living organisms need to live. Farmers use fertilizers that consist of mostly Nitrogen and Phosphorus to help their plants grow. Even though the air is 78% Nitrogen, living organisms can't get their Nitrogen through the air like we get our Oxygen. Nitrogen has to be broken down into a different form before it can be used. We breathe the nitrogen in the air into our lungs and then right back out again without using any of it!

3. Some plants can convert the Nitrogen in the air into a biologically useable form. These types of plants are called 'Nitrogen fixers' and consist of legume plants like clovers, peas, and beans. Nitrogen fixers have a mutualistic, symbiotic relationship with bacteria in their roots. The bacteria are the organisms that actually fix the atmospheric Nitrogen into a usable form. The plant then takes this Nitrogen and provides the bacteria with sugars and protection.

   Nitrogen in the air consists of two Nitrogen atoms triple bonded together. These three bonds take a lot of energy and a special enzyme to break them apart. The bacteria in the roots of the Nitrogen fixing plants can break that tough triple bond and convert the Nitrogen into Ammonia. Since Ammonia has only single bonds with its Hydrogen atoms it is much easier for the plant to use.

$$N_2 + 8 H^+ + 8 e^- \longrightarrow 2 NH_3 + H_2$$

Air Quality #2

Maria leaves the previous room and enters a room that contains 100% oxygen molecules.

Can you explain what effects Maria would experience and the time that these effects would begin and end?

1. The oxygen was in the form of O_3 not O_2 so Maria would have a severe cough caused by high levels of this ozone and would eventually die from a lack of breathable oxygen!

2. The oxygen in this room was in the form of Ozone (O_3). Normal breathable oxygen consists of two oxygen atoms bonded together (O_2). Ozone is three oxygen atoms bonded together and is actually toxic to human health! At ground level ozone is a pollutant that is produced by car exhaust. However, when Ozone is safely above us in the upper atmosphere it creates the Ozone Layer which protects us from harmful ultraviolet sun rays.

3. The chemical structure of Ozone comes in two different forms which we call it's '***resonance forms***.' These structures can be seen below:

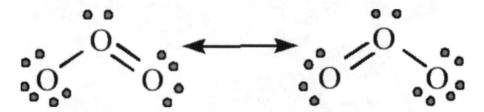

Strangely, chemists have found that both of the bond lengths of each molecule are exactly the same at 1.278Å each. Normally a double bond would pull two atoms closer together than the single bond. This shows us that the resonance structures above are not totally correct but only serve to help us think about how Ozone reacts with other molecules.

To illustrate this concept we draw a dotted line on each side of the center Oxygen to represent half a bond each.

Young Thrower

Bill's 4 year old son picks up a baseball. Imitating his father, he winds up and throws the ball. A radar gun clocks the speed of the ball at 97.3 miles per hour which is as fast a professional pitcher.

Can you determine how Bill's son is able to achieve such a feat?

1. Bill's son is throwing the baseball on a train that is traveling in the same direction that he is throwing. The radar gun is on the ground next to the train!

2. If the train is traveling at 70 miles per hour and Bill's son throws the ball at 27.3 miles per hour then the speed of the ball to an observer with a radar gun standing next to the train would be 97.3 miles per hour. The two speeds are added together to create the total speed of the ball.

3. You actually observe this principle of different speeds relative to a source when you hear a police car travel by on the street. When the police car is approaching your position the sound waves that its siren creates are pushed together. The speed of the car is pushing into the speed of the sound waves. This causes the sound to be a higher pitch. Once the police car passes by you the sound waves from its siren get farther apart because the car is now racing away from the sound waves that the siren is creating. This causes the sound to be a lower pitch.

Let's add some numbers to this example. The speed of sound is 768 mph and the speed of the police car is 68 miles per hour. When you are standing ahead of the police car, the sound waves are being pushed towards you.

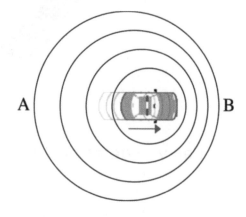

This means that the speed of the sound waves is 768 + 68 = 836 mph relative to where you are standing. When the police car passes you the sound waves are traveling at 768 – 68 =700 mph.

Neutron Star Dust

A space probe is sent to the nearest neutron star. The specially designed craft lands and collects 1 teaspoon of the star's material.

When it arrives back on earth, scientists place the small amount of star on a very accurate scale.

Can you give a close estimate to what this teaspoon amount of material might weigh?

1. One teaspoon of neutron star would weigh about 20 billion pounds! (or 10 million tons!)

2. A neutron star is created by a supernova, a giant explosion of a star like our sun. Average stars, like our sun, are made up of quadrillions of atoms. Atoms are made of protons, neutrons, and electrons. Atoms are actually 99.99% empty space because the electrons of the atom need a lot of space to move around. You can see this in the figure of an atom below. In a neutron star there are no electrons or protons, only neutrons! This means that there is almost no empty space in a neutron star, making it very, very heavy.

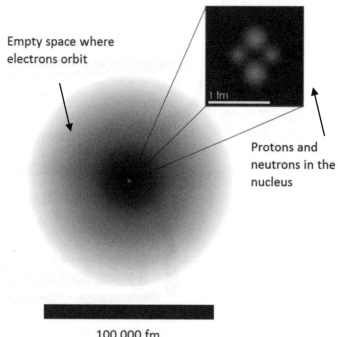

Empty space where electrons orbit

1 fm

Protons and neutrons in the nucleus

100,000 fm

3. Neutron stars are usually around one to two solar masses (the mass of our sun). Even though their mass may be similar to our sun, their size is not. The radius of a neutron star is only around 10 km compared to the sun's 60,000 km radius. Since gravity is dependent on two factors, mass and radius, the gravity of a neutron star is massive. For example, gravity on the surface of the earth is a steady 9.8 m/s^2 while gravity on the surface of a neutron star would be 5,000,000,000,000 m/s^2.

Water Wonder

Bill and Maria decide to conduct an experiment. They each fill a glass with water and place it outside.   The glasses are identical and the water is pure.

It does not rain and no wild animals disturb the water in any way. Yet in the morning, Bill's glass is still nearly full and Maria's glass is nearly empty.

Can you determine what caused the difference?

1. Bill is in Florida with has a very humid climate. Maria is an Arizona which has a very dry climate. Overnight Maria's water evaporated into the air while Bill's stayed right in the cup!

2. In every climate on earth there are at least some water molecules in the air. We call water molecules in the air '**water vapor**' and we measure this with '**relative humidity**.' Relative humidity is the amount of water vapor in the air compared to the maximum amount of water vapor that the air could hold. Climates like rain forests have almost 100% relative humidity meaning the air there is holding as much water vapor as it possibly can. Other climates like deserts are the opposite and have almost no water vapor at all in the air.

3. When Bill's cup sits outside in Florida there are many molecules of water vapor in the air like in the diagram below. These molecules of water are in a gas state, however when they touch the surface of Bill's cup of water they can sometimes turn into a liquid state. Conversely, the water molecules in the water sometimes get excited and can go from a liquid state to a gas state. When the amount of molecules going from the air to the water is equal to the amount of molecules going from the water to air then the system has reached equilibrium and will not change. Factors like heat, movement, and humidity can all affect where this equilibrium point will be.

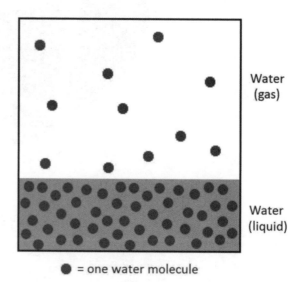

Water (gas)

Water (liquid)

● = one water molecule

Water Wonder #2

Bill and Maria wanted to know how long it would take for a glass of water to freeze.

They each placed a glass of water outside and timed how long it took until the water had frozen solid.

The glasses and water were identical in every way. The outdoor temperature remained exactly the same for the glasses.

Strangely, Maria's water froze faster than Bill's.

Can you determine the reason behind the difference?

1. Maria's glass of water was not protected by the wind which caused the wind chill to cool down her glass faster than Bill's glass!

2. To understand how wind chill works, think about the surface of the water in each glass. The molecules of water in each glass are moving faster than the surrounding air molecules. The faster that a substance moves on a molecular level, the hotter that substance is. Since the water in the glass is moving faster, it heats up the air that is immediately above the water. This creates a layer of slightly warmer air which insulates the water underneath. In Bill's glass, this insulating layer keeps Bill's water at a higher temperature for longer (although even his water will freeze eventually!). Maria's glass of water is in the wind so her insulating layer is constantly being blown away and replaced with colder air!

3. Wind chill is very important to humans because we must keep our body temperatures at a homeostatic level of around 98°. Humans have evolved to sweat in order to increase heat loss in the body. In order to sweat efficiently we need some sort of air flow over our skin in order to move the insulating layer of warm air molecules away from our bodies. Without air movement, such as wind chill, sweating would be useless as in moist, humid climates.

Insulating Layers of Air at Different Wind Speeds

0 mph   10 mph

Water Wonder #3

Bill and Maria place a glass of water outside. The glass was outside all night long. The temperature was a constant 30 degrees. When they retrieve the glass the water is not frozen.   Nothing has been added to the water - the water is pure.

Can you explain why the water is still in liquid form?

1. The outside temperature was thirty degrees Celsius not Fahrenheit!

2. There are many different ways to measure temperature and they are all based off of different reference points. The Fahrenheit scale that we know in the USA is based off of the freezing point of brine water (0^0F), the freezing point of water (32^0F), and the temperature of some humans (96^0F). The Celsius scale is much more widely used and is based on water's freezing point (0^0C) and water's boiling point (100^0). The USA, Belize, and the Cayman Islands are the only countries still using the Fahrenheit system today!

3. Kelvin is the preferred system of measurement for most Scientists. The Kelvin scale has the same temperature increments as the Celsius scale but Kelvin sets the 0^0 mark at absolute zero. Absolute zero is the point where all thermodynamic motion stops in a substance. Absolute zero is impossible to measure with our current technology because any effort to make a measurement of an object at absolute zero will raise the temperature of the object very, very slightly. Quite the paradox!

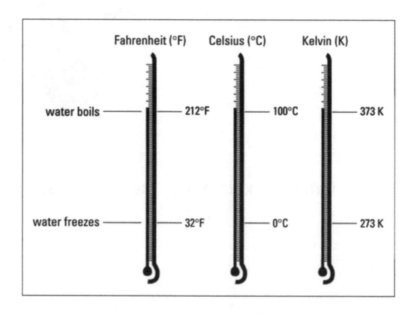

Sonic Booms

Bill is attending an outdoor event.   Suddenly, he begins to hear multiple sonic booms.

There are no planes or rockets in sight.   The sounds are not from any type of explosion.

Can you determine where Bill might be?

1. Bill is at a rodeo! He is hearing the sonic booms from a cracking whip.

2. The loud cracking noise that you can hear when a whip is snapped is caused by the end of the whip traveling faster than the speed of sound. This sound was useful when ranchers needed to move cattle around their land. The cracking noise scared the cattle in the direction that the rancher wanted them to go.

3. The human arm cannot shake a whip faster than the speed of sound so physics is needed to make a whip's sonic boom. The conservation of momentum says that momentum equals the mass of an object times the object's velocity:

$$p = mv$$

When a whip is moved the energy travels from the handle all the way out to the end or 'snapper'. As the energy moves the mass of the whip is decreasing but the momentum stays the same. This means that the velocity the whip moves must increase to compensate for the decrease in mass. When the energy reaches the end of the whip there is almost no mass which means that the velocity must be extremely high. This causes the velocity of the snapper to exceed the velocity sound which creates a sonic boom.

1 + 1 = ?

Maria carefully measures out a quart of water. Bill walks into the room with a quart of another liquid.

They carefully pour them into a third container but strangely find that it doesn't measure up to 2 quarts. Neither Bill nor Maria spilled any liquid and none of either liquid had evaporated.

Can you determine what caused this mystery?

1. Maria's quart of water combined with Bill's quart of alcohol. When alcohol and water are combined the result is less than two quarts of liquid!

2. Molecules of the alcohol Ethanol are much larger than molecules of water. This means that when the two are mixed the water will 'fill in the cracks' of the ethanol. An example that is easier to see is adding water to sand. If you add a half jar of water to a half jar of sand you will not get a whole jar of sand and water! The small water molecules fill in the empty space between the much larger sand molecules.

3. The figure on the right shows water and ethanol molecules in a solution. The water molecules are smaller with only one Oxygen and two Hydrogens (H_2O). Ethanol is bigger with Carbon, Oxygen, and Hydrogen (C_2H_5OH).

Ethanol and water molecules

In chemistry, alcohols always have an Oxygen atom attached to a Hydrogen atom in the molecule. This is called a hydroxyl group. A hydroxyl group is an example of a 'functional group' or a group of atoms that gives a molecule certain characteristics when reacting with other chemicals.

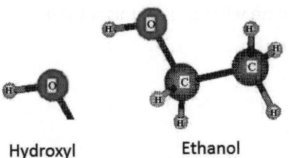

Hydroxyl Group

Ethanol

Seen and Unseen

Scientists wonder what would happen if they took all the observable matter (stars, planets, galaxies, etc.) and put it on one side of a large scale then put all the unseen matter (dark matter) on the other side of the scale.

Which side do you think would weigh more?

1. Dark matter would weigh much more!

2. Dark matter is a thought to be an invisible substance that permeates the universe with even more frequency than regular matter. Since it is invisible, the only way that scientists can predict that dark matter exists is by seeing the effects of its gravity on other objects. This is stellar algebra! The scientists know how much visible matter there is in a galaxy and they can observe how strong the galaxy's gravity is. These two variables don't seem to be equal so the scientists know that there must be at least one other unknown variable: dark matter!

3. From these calculations, scientists have estimated that 23% of all the matter and energy in the universe is dark matter. This compares to only 4.6% that consists of normal matter. It seems that most of the matter in the universe is invisible to humans!

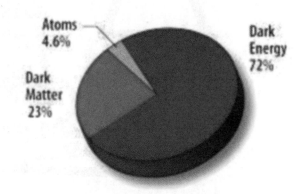

Matter and energy in the universe

This invisible matter is probably made up of particles called WIMPs (weakly interacting massive particles). These particles only interact with 'weak' forces in the universe like gravity. They would be slow moving and therefore cold because heat is merely how fast particles are moving. WIMPs would also be dark because they do not interact with light or the electromagnetic spectrum.

Killer Radiation

Bill and Maria had been observing and studying the sun. Their calculations predicted that in 22 days the sun was going to produce intensely strong and powerful radiation rays. It would be strong enough to burn and kill all living life on planet earth.

Can you determine what Bill and Maria might do in preparation for this day?

1. Bill and Maria won't do anything! Every day these massive amounts of radiation are stopped by the ozone layer before they can hit the Earth's surface.

2. The ozone layer is a layer of ozone gas (O_3) which exists in the stratosphere. When harmful ultraviolet (UV) radiation from the sun enters our atmosphere, it hits the ozone molecules and they absorb the radiation's energy. This means that only a small fraction of harmful UV-B and UV-C rays are able to hit the earth.

3. When UV rays from the sun enter our atmosphere, they hit the ozone (O_3) molecules and break them apart. Breaking the ozone molecule's bonds absorbs the UV rays' energy and turns ozone into a breathable oxygen molecule (O_2) and a lone oxygen atom (O). Our ozone would be destroyed very quickly if this were the end of the process but luckily the lone oxygen atom (O) is not stable and almost immediately binds itself back to a breathable oxygen molecule (O_2) to again make ozone (O_3).

When ozone is in the bottom layer of the atmosphere, the troposphere, it is actually bad for our health. Cars and factories give off pollutants like ozone as waste protects. When these pollutants build up, usually in the summer, they create what we know as smog.

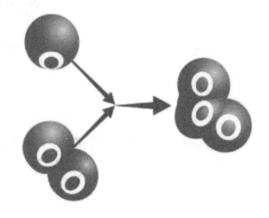

The Chicken or the Egg?

Maria and Bill decided to discuss the age old question of which came first, the chicken or the egg.   Being scientists, within a few minutes of discussion they came to a solid conclusion.

Can you determine what their conclusion was?

1. The egg came first! Reptiles laid eggs for millions of years before chickens had even evolved yet.

2. Scientists can see from fossil evidence that the first reptiles evolved in the Pennsylvanian period of time which was about 340 million years ago. Most reptiles laid eggs in order to reproduce. Eventually one type of reptile, most likely the Archaeopteryx, branched off from other reptiles and became the ancestor to all known bird species!

3. Although the exact sequence of bird evolution is debated, most all scientists agree that birds are the descendents of reptilian dinosaurs. Birds and dinosaurs share many common characteristics. Multiple species of dinosaur fossils have been found with feathers. Dinosaurs have also been found to have hollow bones. These bones are pneumatized which means that they have air pockets inside them to make them lighter.

One last interesting similarity between dinosaurs and birds is they both seem to have gastrolithes to aid in digestion. Gastrolithes are rocks swallowed into the stomach to help grind food in mechanical digestion. Modern day birds have a separate organ to hold these 'stomach rocks' called a gizzard.

Which Ship?

Spaceship A and Spaceship B are on the moon ready to take off for the journey back to earth. Spaceship A is the shape of a cube. Spaceship B is in the shape of a typical rocket or pencil.

Both ships weigh exactly the same and have the same exact engines. The ships take off at the same time.

After 10 minutes of acceleration, can you determine which ship would be closer to earth?

1. Both ships are the exact same distance away from the Earth!

2. Since there is no air in space there is also no air resistance. With no air resistance the shape of an object does not relate to its speed! In the Earth's atmosphere, the cube shaped ship would move much slower than the pencil shaped ship because the cube shaped ship would hit the air more directly. A classic example of this principle is dropping a bowling ball and a feather at the same time from the same height. In the atmosphere, the bowling ball will hit the ground first but in a vacuum (no air) they will both hit the ground at the same time!

3. The force that air puts on a moving object is called drag. Since spaceships A and B are in space they have no drag. However, if they were in Earth's atmosphere the force of air atoms hitting the hull of the ship would be a force in the opposite direction of the ship's motion.

   Olympic swimmers give us a good example of drag. In order to achieve maximum efficiency in the water they need to reduce the amount of drag on their bodies. They do this by reducing the surface area of their body in the direction of the water flow. This causes less water molecules to hit their body and therefore reduces the amount of drag force pushing them backwards.

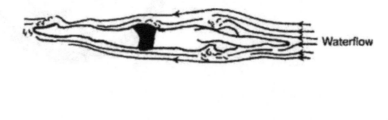

Pencil Puzzler

Bill looked down at the pencil on his desk. Upon closer examination, he noticed it was moving. His desk was level. There was no wind and no earthquakes or any movement of his desk.

Furthermore, his desk is at his home and the pencil is not moving because the earth, solar system, galaxy etc. is moving.

Can you determine how and why his pencil is moving?

1. Anything that has a temperature above absolute zero is moving microscopically!

2. On a molecular level, having heat is the same as having motion. The more heat energy an atom has the faster it will vibrate and move. A pencil resting on a table is probably about room temperature at 23⁰ Celsius (about 72⁰ Fahrenheit). This means that its atoms are moving very quickly compared to absolute zero which is -273.15⁰ Celsius or -459.67⁰ Fahrenheit. Now that's cold!

3. Many scientists have tried to get different substances to absolute zero. However, this is a very difficult task. To start, atoms will absorb any energy that they can from their surroundings. In order to get an atom or molecule to absolute zero a scientist would need to separate the target substance from any surrounding atoms.

Another problem is observing the atom or molecule at absolute zero. A scientist must use a detector that uses light or electrons to bounce off of the substance in order to measure the temperature. If the scientist has a substance at one billionth of one degree Kelvin and then hits that substance with light to measure the temperature, then the temperature will increase! The atoms will be excited by the measuring tool and absorb the energy that it contains.

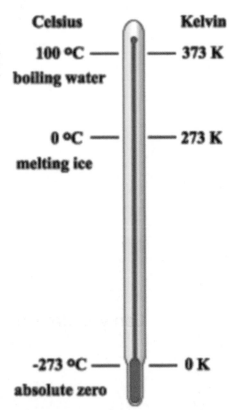

Beach Wave

Bill and Maria were enjoying a nice day at the beach. Suddenly, they were hit by a wave traveling at 760 miles per hour! Amazingly, they survived and continued to play in the water.

Can you explain this puzzling scenario?

1. Those waves that hit Bill and Maria were sound waves!

2. Sound waves are simply vibrations of the air between an object moving and another object, like an ear. Different types of sounds are caused by the different frequencies of these air vibrations. Our human ears can detect (hear) any sound wave between the frequencies of 20 and 20,000 Hz. As humans grow older their ability to hear sounds in the high pitch, upper frequency range is diminished.

3. Waves come in two different forms; longitudinal and transverse. Transverse waves move up and down as they travel similar to ocean waves. Longitudinal waves move side to side like a slinky. Sound waves are longitudinal because they are formed by the compression of air molecules. This compression of air forms alternating areas of more dense and less dense air. These differences in densities create differences in pressures which a human ear can detect.

   The less dense areas of the wave are the areas of rarefaction. The measurement from the beginning of one rarefaction zone to the beginning of the next rarefaction zone is one wavelength of sound.

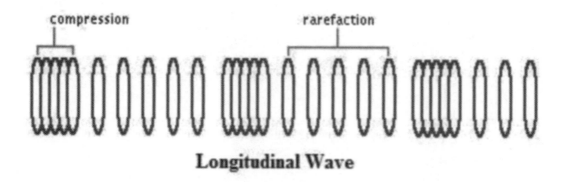

Longitudinal Wave

Deadly Hit

A body was brought to the coroner. The only information given was that whatever hit the man ...killed him.

Strangely, the coroner found no broken bones, bruises, scratches, or bleeding on the exterior of the body.

Can you determine what hit and killed this man?

1. A high energy sound wave killed this man!

2. Scientific researchers have found that the human body is very sensitive to high decibel sounds. Human ears will start to hurt around the 130 decibels range. This sound level is similar to a jack-hammer or gun shot. A jet engine puts out around 150 decibels. Right above this level is when the human eardrum starts to rupture. The higher the decibel count goes, the more damage the human body takes.

3. Scientists are not exactly sure what will happen to the human body when the decibel range reaches higher than 150. In a study of mice, scientists found that at 185 decibels the mice's livers and lungs took heavy damage. The most damage to the body seems to occur in the organs that have empty cavities. The stomach and intestines are also highly affected by intense sounds. When these sounds have been used on crowds of protesters the result was many people vomiting and getting severe intestinal pain.

Baffling Blindness

Bill was abducted. He was taken 2 miles away from his home. When he awoke, he found that he had been blindfolded. Try as he might he could not remove the hood. Yet, in the <u>middle of the night</u> the captors accidently left the door open and Bill safely made his way home.

Can you determine how he performed this difficult task?

1. Bill is a bat and uses echolocation to find his way home in the dark!

2. A bat can send out pulses of sound through its mouth that hit the objects around them. These sound waves then bounce back to the bat's ears where they can be decoded into the object's location. When bats are flying they put out an average of 10-20 sound pulses per second. This may sound like a lot but when a bat gets close to its insect prey, it will start to emit over 200 sound pulses per second!

3. One main food source for bats that use echolocation is Lepidoptera (moths). Lepidoptera have evolved many different defense strategies to throw off the bat's targeting system which can be seen below:

   Acoustic Concealment – Moths fly close to the ground to clutter and confuse the echolocation of the bats.

   Mimicry – Some moths can emit sounds which mimic and confuse bats that use echolocation

   Physical Avoidance – Moths will stop flying when they hear a bats echolocation. They drop out of the air to avoid being eaten.

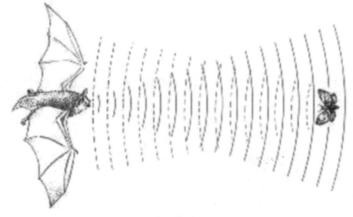

Twins Swim

Two identical brothers decided to go for a swim. They went to a park that had side by side lakes. One brother went to Lake A and found it hard to float. The other brother went to Lake B and found it hard to sink beneath the surface.

Both brothers weighed the same and were wearing identical swim trunks - nothing else.

Can you determine what's going on?

1. Lake B is much saltier than the fresh water lake A.

2. Even though the two twins have the same mass, they can float differently in the two lakes due to the amount of water that they displace. The twin in lake B will displace less water because the salt water is heavier and can hold him up more easily. The twin in lake A will sink easier because he is heavier when compared to the lighter fresh water. If you were to take a bath in heavy Mercury like Gengis Khan used to then you would float very, very easily. In fact, it would be hard to stay in the liquid!

3. Buoyancy is the ability of an object to float in a substance. The ability to float all comes down to density. The human body is made up of almost 80% water and is therefore very similar in density to water. Parts of the human body would sink while others would float. For instance, fat is an oil and oil floats on water. Bones and muscles are densely packed and therefore sink. However, if you really want to change your density so you can float in the water then take a deep breath! Air is one of the least dense substances on the planet and can even make rocks float, like in the case of Pumice.

Golden Air

Bill bought what he thought was a small bar of solid gold. Yet, Maria examined it and told him that what he had purchased was mostly empty space.

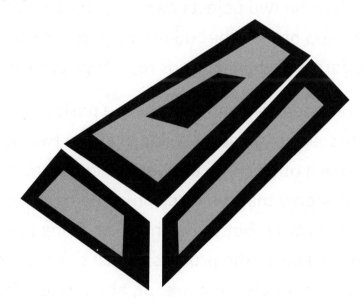

Can you determine how the seller was able
to fool Bill into buying the gold?

1. Almost all matter on Earth is mostly empty space, including gold!

2. One atom of gold consists of 79 protons and 120 neutrons held in an inner nucleus and 79 electrons orbiting this nucleus in fixes orbitals. The space in between the electrons and the nucleus turns out to be about 99.99999999% of the entire atom's volume. Humans can't detect this empty space because when one object makes contact with another object the electrons in the outermost shells of the atoms repel each other. In fact, no two objects ever truly touch. One object is floating on top of the other object very, very closely due to the electrical forces of the electrons repelling each other.

3. Electrons are negative and protons are positive so why doesn't the empty space of an atom collapse as the two particles attract to each other? The answer lies in the tremendous amount of energy that the electron contains. Electrons are attracted to protons but they tend to go flying past the nucleus instead of sticking to the protons. Since electrons have a certain speed they always travel at, about one third the speed of light, they cannot slow down enough to join the nucleus. When they go flying out the other side of the nucleus they are slowly pulled back again which starts the process all over again. Therefore, the electron will always be flying around the nucleus within a certain distance as long as no other forces are acting on the atom.

Water Worm

Maria decided to conduct an experiment while on the space station. She carefully squeezed some water out of a container into a long line.   While it slowly floated, it reminded her of a worm.   She left it suspended there as went to bed.

Can you determine what the water might look like in the morning?

1. By the morning the "water worm" would look like a sphere!

2. Water molecules will naturally be attracted to each other due to their cohesion. This means that there is a force bonding water molecules together. If multiple water molecules come near each other then they would naturally stick together rather than bump into each other. An example of this would be a glass just barely too full of water. Since water molecules stick together, it is possible to fill a glass over the rim and still keep all of the water in the glass!

3. Each water molecule can connect with four other water molecules by making Hydrogen Bonds. The Oxygen in a water molecule pulls on its electrons so hard that the Oxygen takes a slight negative charge (electrons are negatively charged). This means that the Hydrogen atoms will have a slightly positive charge because of their lack of electrons. These positive and negative charges attract other water molecules that also have positive and negative charges. Since each water molecule has four places to bond with other water molecules, scientists call this a tetrahedral shape.

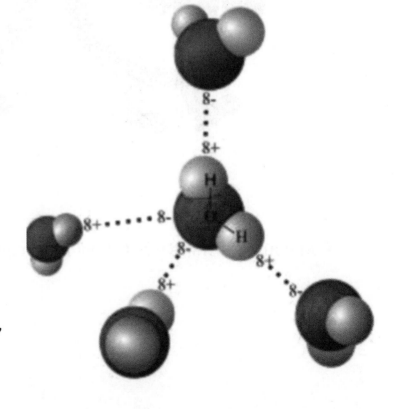

Straight as an Arrow

Bill put on a space suit and left the shuttle. He took out a bow and pulled back a specially designed arrow that sent back data on its path and speed. After one year, they noticed that the arrows' path was curving and its speed increasing.

Can you explain this situation?

1. The arrow was passing by a star and being pulled by the star's massive gravity!

2. When an arrow gets caught in the gravity of a large space object, it gets a new force pulling on it. This new force slowly works to pull the arrow in the direction of the star therefore bending the arrow's path. If the arrow is travelling fast enough then it can actually escape the gravitational pull of the star and be hurled back out into space in a new direction, similar to how you would be thrown off of a merry-go-round if you were not holding on!

3. Since this idea of a '**gravity assist**' was developed in 1961 it has become a common practice in order to save fuel on all different types of spacecraft. For example, NASA's Voyager probes used the gravity of Jupiter, Saturn, and Uranus to become the first manmade objects sent outside of our solar system!

A gravity assist works by sending the spacecraft right behind the path of a large space object like a planet. The planet will then pull the spacecraft in the direction that it is moving, therefore making the spacecraft increase its velocity. The opposite effect can also benefit space travelers. If a spacecraft is sent to the opposite side of the planet, then its velocity can be slowed down.

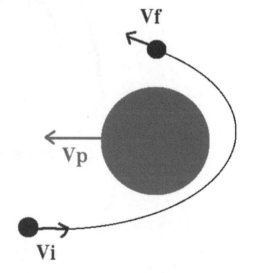

Ghostly Image

Bill and Maria have created a machine that transports a human from one spot to another. The machine was designed to only work on human cells.

Their first true test seemed to be successful. The human test participant, Bill, vanished from spot A and reappeared in spot B.

Yet, strangely, there was a ghostly white outline of Bill left behind. It quickly fell to the ground.

Can you determine what this pile of white substance was?

1. The ghost-like white outline was all of the bacteria that lived on and in Bill's body being left behind!

2. Each human body has its own ecosystem of 'microbiota' or tiny microbes that call your body home. In fact, there are around ten microbial cells for every one human cell in your body! Luckily these microbial cells are much smaller than our body's cells.

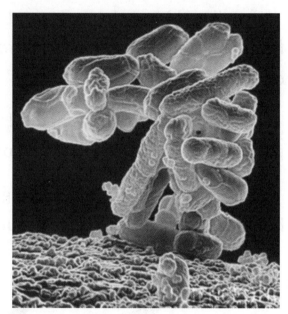

   The picture on the right shows a group of E. coli bacteria cells that live in human colons. Most E. coli are actually beneficial to the body but some mutant strains do cause disease.

3. Most microbial life in the human body is commensal. This means that the microbes and humans both benefit from the relationship. Intestinal bacteria are a good example. The bacteria break down carbohydrates that the human gut cannot naturally break down. In exchange for this service, the bacteria is provided with a constant source of food from the human.

   Some microbial life in the body can actually cause disease. About thirty percent of all humans have Staphylococcus aureus living in their nasal cavity (nose). Staph bacteria cause infections of the skin. Luckily Staph is usually not a threat as long as your immune system is strong and your good bacteria are plentiful!

Twinkle, Twinkle...

Maria was enjoying a clear night. As she gazed at the stars, she noticed that some were twinkling and others were not.

Can you explain this difference?

1. Stars twinkle but planets in our solar system will not!

2. Stars are so far away from Earth that they seem to just be tiny dots of light. These dots of light have to pass through Earth's atmosphere to reach your eye. When light passes through the atmosphere it refracts or bends in different ways. This constant refraction of the light in slightly different directions is what causes the twinkling effect. In space, astronauts do not see any twinkling stars!

3. Light refracts when it travels through mediums with different densities. Different levels of the atmosphere have different temperatures and densities of air. Since air in the atmosphere is moving, these differences in density can disrupt the light's path many times before it hits the human eye.

A star in the sky is one small point of light hitting our eye. This means that it is greatly affected by any disturbance that the atmosphere creates. Even though planets look just as small to the naked eye, they are actually slightly larger disks rather than points. This means that the disturbance by the atmosphere on the light of the planets is more balanced out, causing almost no twinkling.

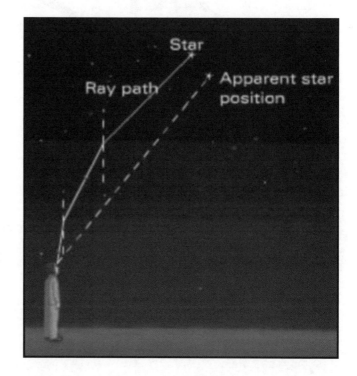

Twinkle #2

As Maria further examined the night sky she noticed that some of the stars (the points of light that were twinkling) were of different colors. Some were redder in color, some more blue, some more yellow, and some more white.

Can you conclude if this was due to size, distance, temperature, or composition of the star?

1. The color of a star is directly determined by its temperature!

2. The thermal energy (temperature) of a star is determined by how much mass the star has. The more mass that the star has the more nuclear fusion will happen in its core. Nuclear fusion is a process where matter is converted into energy. This conversion creates a massive amount of heat which can change the color of a star! As the star gets hotter the light emitted moves more and more towards the blue end of the color spectrum.

3. Incandescence is the conversion of heat energy into light energy. All objects are incandescent to a certain degree but most objects emit too little light to detect. One example of incandescence that we can see would be hot iron. Iron starts out 'orange hot' and as the temperature increases the color shifts more toward the smaller wavelengths of the electromagnetic spectrum. Stars behave in the same manner. A star of about 5800K, like our sun, will emit yellow light while a star of about 28000K, like the blue dwarf Bellatrix, will emit blue light.

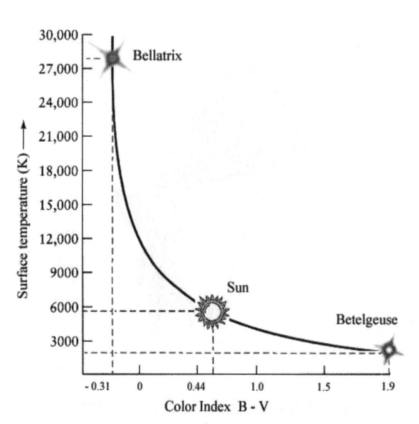

Miracle Walk

One day, Bill tells a classroom of students that he can walk on water. They are accustomed to his stories and many students raise their hand to say things like:

1. When the sidewalk is wet...we walk on water.
2. There is water underneath most of the surface of the earth, so in a sense, we are walking "on water."

Bill smiled and then took the class outside to a nearby lake and, to the amazement of many, walked out to the middle of the lake and back to shore. He remained on the surface at all times. Other than his shoes, there was nothing between him and the water.

Can you conclude how he was able to pull of this miracle?

1. Bill was walking on the ice of a frozen lake!

2. Most substances become denser as they decrease in temperature because their atoms are not moving as quickly so those atoms can pack in more closely together. Water behaves in the same way until it reaches its densest point at 4°C. Then the hydrogen bonds between the water molecules start to get stuck in place. This causes the water to become less dense and more rigid at the same time, creating ice!

3. Hydrogen bonds are weaker than covalent and ionic bonds. However, each water molecule creates four different hydrogen bonds with the water molecules around it. This adds up to a strong force when all of the water molecules are slowed by loss of thermal energy.

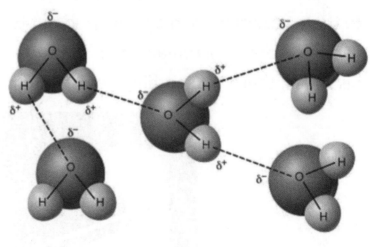

Hydrogen bonds in water are caused by the high electronegativity of Oxygen. The Oxygen atom pulls the electrons of the two Hydrogen atoms closer to its nucleus. This means that the Oxygen atom has a slightly negative charge. Since the Hydrogen atoms' electrons are further from their nuclei, that means that the Hydrogen atoms have a slightly positive charge. The positive charge of the Hydrogen atoms on one water molecule and the negative charge of the Oxygen atom on another water molecule bond to make a hydrogen bond!

Weird Walk

Maria wakes up one morning and decides to take her dog for a walk. She and Newton leave the house and begin to walk due East the entire trip. After two kilometers of walking they find themselves back where they started.

Can you explain this paradox?

1. Maria is walking her dog only a few hundred feet from the North Pole. The circumference of the Earth is much smaller there!

2. The circumference of a sphere is the distance you must travel to get around the object. Since Maria was closer to the top of the Earth, the circumference was very small relative to the Earth's axis. The largest circumference on Earth relative to the Earth's axis is at the equator. The equator is about 40,075km around! If Maria started walking around the equator at a normal walking pace of 5km/hr she would be back where she started in a little less than a year... if she never stopped walking!

3. The mathematical formula for circumference is as follows:

$$\text{Circumference} = (\text{Diameter}) \times (\pi)$$

This equation is one of the most basic uses of the mathematical constant pi (π). Pi is an 'irrational' number which means that it cannot be expressed as a ratio of two integers. In other words, no known fraction of integers works out to be exactly pi.

3.141592653589793238462643
3832795028841971693993751
0582097494459230781640628
6208998628034825342117067
9821480865132823066470938
4460955058223172535940812
8481117450284102701938521
1055964462294895493038 19
6442881097566593344612847

Pole Puzzler

Bill makes a phone call to Maria and says, "I made it to the North Pole!"

Maria replies, "Uh..so did I. I am standing there now but, I don't see you."

Bill looks around and sees no one in sight and there is nothing to hide behind.

Bill and Maria are left scratching their heads.

Can you help them solve this mystery?

1. There are two different north poles on Earth, one is magnetic and one is geographic!

2. The geographic North Pole is the top of the axis on which the Earth is spinning. This is the pole that is in the arctic. Robert Peary was the first known man to reach this point in 1906. The magnetic North Pole is actually in the Canadian Arctic Islands. When your compass points to the North, that is where you are actually headed!

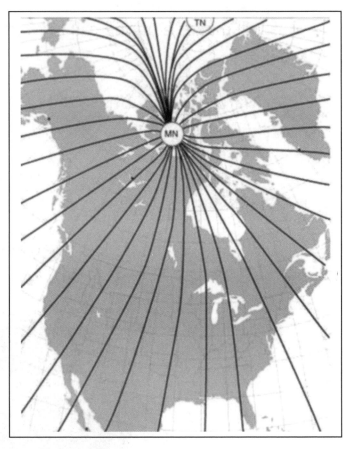

3. The North end of a compass points toward the magnetic North Pole. This means the magnetic North Pole is actually the South end of the Earth's magnetic field because opposite magnetic poles attract.

The magnetic poles on Earth are caused by the constant moving of the Earth's mantle and core. As the molten iron and rock move, so does the magnetic pole. Over the last four hundred years the pole has moved hundreds of kilometers in seemingly random directions.

Sometimes this pole movement can be even more dramatic. Evidence suggests that the whole polarity of the Earth completely reverses every 300,000 years or so. Scientists can measure this flip in direction of the Earth's magnetic field by looking at ancient magma samples. As the magma cools it aligns in the direction of the Earth's magnetic field. Our last magnetic field flip was about 800,000 years ago so we are well overdue!

Dog Bite

An angry dog is chained to a tree. The chain is 10 feet long. Bill walks 15 feet away from the dog, yet he is still attacked and bitten.

Can you determine how this could have happened?

1. Bill is 5 feet away from the tree on one side while the dog is 10 feet away on the other!   That totals 15 feet.

10 feet ——— 5 feet

2. There are numerous examples in science discoveries that remind us how simple it is to miss the obvious. Scientists must continually be conscious of all the possibilities of an experiment. Most scientific discoveries throughout history start with a scientist looking at a totally normal object or phenomenon from a different perspective!

3. Although this is a simple math scenario, math IS a significant part of science. Often it is the math that becomes the most significant part of a science discovery.   For instance, most scientific discoveries in Physics actually work backwards from other scientific disciplines. In Physics, usually a mathematician first shows that a concept is mathematically possible. Only then does a Physicist actually test that concept with an experiment. This is how many Physics discoveries were made including the existence of protons, neutrons and electrons and also Albert Einstein's famous $E=mc^2$ equation!

Water Crisis

Bill walks into a room and sees a bottle of water 12 feet away. He is thirsty yet is unable to get his hands on the bottle. The bottle is not behind anything. There is only air between Bill and the bottle.

Can you explain this dilemma?

1. The water is hanging 12 feet above him!

2. Quite often, people tend to think in only two dimensions.   They assume that this is a 'usual' room and a 'usual' placement of a water bottle.   However, much of the time in scientific discoveries the answer is not 'usual' at all! That is why everybody doesn't see the answer right away. Scientists must always consider geometry and dimensions in their work.

3. A good example of this is velocity. Velocity cannot be measured in the same way as speed. Speed is the distance an object has travelled in a certain amount of time. We call this a 'scalar' quantity. Velocity is exactly the same except with a <u>direction</u> added on. Adding a direction to a speed makes velocity a 'vector' quantity. Vectors take into account both direction and magnitude.

So why can't we just use speed? Why do we need numbers to be vectors? Well, in the real world objects do not always move in a straight line. This means that without a vector, calculating an object's location would be very difficult. For example, a car goes 30 km/hr from a house. After two hours where is the car? Sixty kilometers away? Only if the car didn't make any turns! And in which direction is the car now? There is no way to know with just a scalar quantity. A vector is needed to add direction.

Boil Buster #1

Maria fills two pots with water. The pots are of the same dimensions and same weight and she fills them with the same amount of water. She places them on the stove.   The heating elements are heating at exactly the same rate.   She notices that one pot begins to boil before the other.

Can you determine what the difference is?

1. The pots are made out of different material! One is made of copper and the other is made of stainless steel.

2. Copper and stainless steel are substances with different heat capacities. This means that the rate of heat that can be transferred between the heating element and the two pots will be different. The copper pot, which has a high heat capacity, conducts heat to the water inside much more quickly than the stainless steel pot because stainless steel has a much lower heat capacity in comparison. This is the same property that is used in coolers to keep items cold. The material used to make coolers has a low heat capacity and therefore does not let heat in from the environment very quickly.

3. A substance's heat capacity is determined by the specific heat value of the substance. Specific heat is the amount of energy a substance will intake before there is a resulting change in temperature. For instance, the copper and stainless steel pots had the same amount of energy inputted into the system but that energy moved the molecules of copper much easier than the molecules in the stainless steel. This increased movement could be a result of multiple intermolecular properties but in this case it is probably a result of an increased amount of electron movement in the copper. Since there are more atoms and electrons in the copper to accept the energy, that energy can be transferred more quickly.

Boil Buster #2

Maria fills two copper pots up with the exact same amount of water.   She heats them on stove.   Again, one boils faster than the other.

Can you conclude what is now affecting the boiling rate?

1. Maria added a stone to one pot in order to create a place where air bubbles could form!

2. Water boiling is a result of added heat which makes molecules of water move more rapidly. When these molecules move faster they create more space around them. In a perfectly smooth container this increased space will spread throughout the liquid which increases the time until bubbles are created. If the surface in the pot is NOT smooth, like when you add a stone, then the bubbles that result from the increased space in between the water molecules will be able to form more easily. The uneven surfaces create a place for the bubbles to congregate!

3. Without an uneven surface for bubbles to form, a substance can actually be heated much higher than its normal boiling point. These situations result in a superheated substance that has the potential to release all of its vaporization energy at once. As soon as an irregularly shaped (not smooth) object is added, the bubbles have a place to form, or nucleate, and all of the energy in the substance will be released at once in the form of superheated vapor.

   This superheating can only occur because of the additional force of surface tension on the liquid. The surface tension is creating a force inwards that prevents the liquid from spreading out to the required amount to start the nucleation of bubbles. An uneven surface provides microscopic bubbles that act as a starting point for nucleation to occur. Once the nucleation process starts, there is a chain reaction as the bubbles extend outwards from their original nucleation point!

Boil Buster #3

Maria and Bill decide to do one more boiling experiment.   They make sure the pots, the water, and the heating elements are exactly the same. Nothing is added to the water.   They are taking accurate temperature readings of the water and confirm that they are heating at exactly the same rate.   Yet, Bill's pot begins to boil, and then boils furiously, before Maria's.

Can you conclude what is going on now?

1. Bill and Maria are at two different elevations!

2. Changing the elevation of an object also changes the amount of pressure that is exerted on that object. If Bill is on the top of a mountain and Maria is at sea-level, then Bill's pot will boil more quickly because there is less atmospheric pressure from the air pushing down on the water in Bill's pot. Since pressure and temperature are directly related, if you decrease the pressure on a liquid then you also decrease the temperature at which the liquid boils.

3. Pressure and temperature are related using the Ideal Gas Law or PV=nRT. Let's take a look at what each variable means to get a better idea of how this equation works.

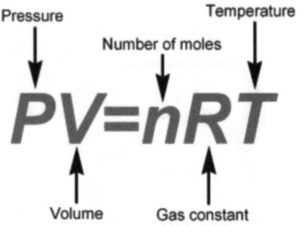

If pressure is decreased on the left side of the equation then another variable must change in order for the two sides to be equal. Volume will stay constant because the overall volume of water is identical in both places. The same goes for number of moles of water. Moles measure the number of molecules of water which is identical in our situation. 'R' is a constant and does not change so only temperature is left to be changed! Since pressure is decreasing, the temperature will also have to decrease in order to balance the equation. This lowers the effective boiling point of the water!

Planet Problem

Maria and Bill were stunned to hear that our solar system had lost a planet.   They knew that for years our solar system had nine planets.   Now, they would only be able to study 8 planets.

Can you determine what scientific event took place
that caused us to lose a planet?

1. In 2006, Pluto was redefined and designated a 'dwarf planet' rather than a normal planet!

2. The area in the general orbit of Pluto is known as the Kuiper Belt. The Kuiper Belt is similar to the asteroid belt but it is more spacious and more massive. In 2005, an object similar in size to Pluto was discovered in the Kuiper Belt. Many other similar sized objects soon followed. Rather than designate all of these new objects as planets, the scientific community decided they had properties more like large asteroids.

3. The evidence against Pluto's planet status has piled up in recent years with more powerful telescopes giving scientists better data about the universe. For instance, it turns out the Pluto's orbit is tilted by 17 degrees from all other planets in the solar system. Also, over 70,000 other objects have been found in the same region of the Kuiper Belt where Pluto resides. All of these objects cannot be designated as planets!

Perhaps the number one reason that Pluto is not a planet is due to its type of rotation. Pluto's center of gravity is not in its center like all other planets in the solar system. Instead, Pluto's moons, especially the largest Charon, cause a barycenter to form. This means the Pluto's center of gravity and rotation is located about 2000 kilometers above the planet's surface. Pluto and its moons orbit each other, not just Pluto!

Steel Stumper

Bill and Maria both had one pound of steel.   Bill places his steel in water and it sank. Maria placed hers in the same water and it floated.

Can you determine why they had different results?

1. Maria shaped her steel into the shape of a boat while Bill's steel was in the shape of a sphere!

2. By shaping her steel into the shape of a boat Maria had created a way to give her steel more buoyancy than Bills steel. Buoyancy is the upward force on an object when it is submersed in a liquid or gas. The buoyancy force on the steel is equal to the weight of the water that is displaced by the steel ball. The more water displaced, the higher the buoyant force! The shape of Maria's steel boat causes much more water to be displaced than Bill's steel sphere because Maria's boat contains air as well as steel below the water's surface, just like in the hull of a boat.

3. Every object on earth has a buoyant force pushing it upwards, even humans! The air in the atmosphere pushes up on the human body with a force based off of the mass of the air. Since air doesn't have much mass, we don't usually feel this buoyant force but we can calculate its strength. The first step of calculation is to find the volume of a human body. On average, a human body is $85,000 cm^3$ or $.085 m^3$.

Buoyancy (B) = Density of Air (p) x Gravity (g) x Volume(V)

$$B = (1.29 kg/m^3) \times (9.8 m/s^2) \times (.085 m^3)$$

$$B = 1.07 kg\ m/s^2 = 1.07\ Newtons$$

So each human has a force of 1.07 pushing them upwards at every moment of everyday. This force may not be very much compared to a dense human but less dense objects like balloons can be greatly affected. Since the density of helium in a balloon is so low, the balloon actually weighs less than the air around it. The upward buoyant force is greater than the force of gravity!

Dying Dilemma

One day, Bill came to work and shared some sad news with Maria. He told her that he had found out that he had a condition where the cells of his liver were dying. She was stunned to find out that in 400 days, every cell in his liver would be dead.

Yet, after 800 days he was still coming to work. When Maria said she didn't like him lying to her, he said he hadn't lied.

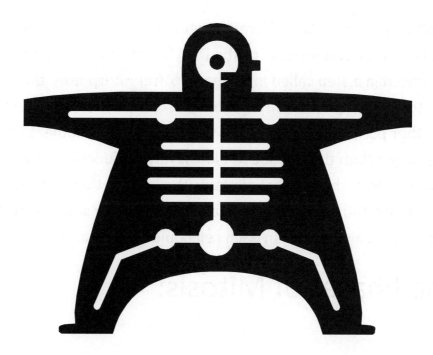

Can you explain this argument and how Bill was able to live?

1. New liver cells were replicating to replace the old cells that were dying!

2. Your body started off as only one cell. One cell divided into two cells, two into four, four into eight, and so on until today when you have roughly 100 trillion cells in your entire body. Cells act like mini machines that are all working together to keep you alive. These machines have parts called 'organelles' which perform functions to keep you alive. Just like a machine, these parts can break down or malfunction and need to be replaced. Replicating new cells to replace the old ones is called '**Mitosis**.'

3. Mitosis consists of four different stages that are listed below. Even before Mitosis can start, all of the DNA in the cell is almost perfectly duplicated in a step called interphase. After interphase, the main goal of mitosis is to split the chromosomes exactly in half, flawlessly. All of the other parts and organelles of the cell are replaceable, but if the DNA is damaged then the cell doesn't know how to function. The DNA is the master code of the cell that contains the information on how to build and run the entire cell. One mistake in copying or separating the three billion base pairs of DNA in each cell could spell disaster!

The Phases of Mitosis:

| Interphase | Prophase | Metaphase | Anaphase | Telophase |

Fast Track Trick

Maria and Bill have built four tracks.   They place a ball at point A and release it.

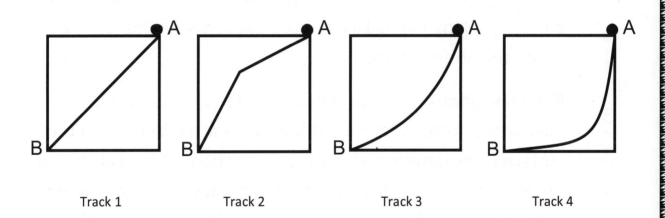

| Track 1 | Track 2 | Track 3 | Track 4 |

Can you determine which track would get the ball
to point B the quickest?

1. Track number four will get to point B the fastest!

2. Which ball gets down the ramp the fastest is dependent on two related factors: **velocity** and **acceleration**. **Velocity** is how fast something moves in a certain amount of time. Whichever ball has the highest average velocity from point A to point B will be the ball to get to the bottom of the ramp first! The best way to gain velocity in the shortest amount of time is to drop the ball as vertically as possible. This lets the ball pick up velocity quickly and allows the ball to keep that high velocity in the more horizontal bottom part of the ramp.

3. This change in velocity over time is called **acceleration**. The acceleration of the ball down the ramp is determined by the force of gravity pulling the ball downwards and the force of the surface pushing the ball away from the surface. This force is called the normal force.

    The gravitational force on the ball in all four situations is exactly the same. That means that the normal force from the surface is the determiner of the acceleration of the ball. Ramp number four has the least amount of normal force decreasing the effect of gravity at the start of the ball's trajectory. This allows the ball to increase in speed more quickly and therefore increases the total average acceleration of the ball over the entire ramp.

Bill's Brick Bridge

Bill and Maria took a break from studying marine life at a nearby beach.   They sat down by a pile of seven bricks. Bill was able to construct an archway with the bricks and pass Maria a sandwich underneath it. Maria was quite impressed.

Even more amazing was the fact that Bill had placed each brick one at a time using only ONE hand.   Bill used no cement or glue to hold the bricks together.

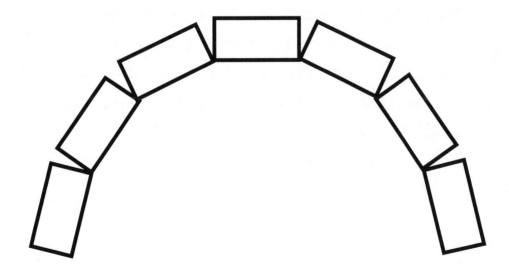

Can you determine how Bill was able to build
this bridge without it collapsing during construction?

1. Bill built his arch on a mound of sand and then took away the sand when he was finished!

2. An arch structure can support itself when construction is complete but when the arch is being built it does not have a keystone to balance the horizontal forces on the stones of the arch. The keystone is the very last stone to be added to an arch and the arch cannot bear any weight until that stone is in place.

3. An arch is able to stand by distributing the forces from the top of the arch all the way down into the ground. This allows the structure to carry much more weight with more stability. If a simple horizontal top were put over a doorway instead of an arch then the weight above door would put a large amount of pressure on the horizontal door frame. If an arch were used for the same doorway then the force of the weight above the doorway would be passed on from stone to stone until it was finally pushed into the ground. This distribution of force is the reason that the arch is much stronger than a traditional horizontal doorframe.

Big Dog Dilemma

Bill and Maria took a 50 pound dog and ran it through their new 'enlarging' machine. The dog came out twice as large in EVERY dimension. The machine did not change the density of bones and muscles.

Can you determine how much the dog would now weigh?

1. The dog would weigh 400 pounds!

2. The reason that the dog's weight would not be one hundred pounds has to do with the changes in volume that occur when an organism grows. Volume is measured in three dimensions (height x width x length) and therefore increases at a much higher rate than one or two dimensional measurements. Even though they are not the same, smaller versions of objects can be useful to scientists. In engineering the idea of 'similitude' is used to test smaller objects instead of larger ones. To test a Boeing 777 an engineer does not start by making a full scale model. Instead that engineer would create a smaller version of the plane that has the same properties but which is much easier to conduct experiments on!

3. A great example of the differences similitude can cause is shown by the surface area to volume ratio of different organisms. Most smaller organisms have a very high surface area to volume ratio which means that they have a large amount of their body touching the outside environment compared to their volume. This allows them to live and grow more efficiently because they are able to dispose of wastes and take in nutrients more quickly from their environment. Large organisms, like a rhinoceros for example, are much less efficient because they have a very large volume to support with the surface area of their body. If a rhinoceros' big, voluminous body is very hot it does not have enough surface area compared to its large volume to cool down!

Tower Trouble

Bill decided to build a tower. He started with a block that was one foot by one foot by one foot (one foot cubed).   He then placed a block on top of that block that was half the size in dimension of the first block.   With a 500 foot step ladder beside him, he continued this pattern of building.

Can you make a close estimate of how his tower would be if he followed this pattern and made a tower of 1 million blocks?

1. Bill's tower will be just under two feet tall!

2. Each time that Bill adds a block that block is half as tall as the previous block. This means that every time a block is added half of the distance between the top of the tower and two feet will be taken up. If only half of the remaining distance is added each time then the tower will NEVER reach two feet in height!

3. The table below shows the math of how the tower is increasing in size but will never reach two feet:

Block Size (ft)	Height of the tower (ft)	Height remaining until two feet (ft)
1	1	1
0.5	1.5	0.5
0.25	1.75	0.25
0.125	1.875	0.125
0.0625	1.9375	0.0625
0.03125	1.96875	0.03125
0.015625	1.984375	0.015625

The pattern in which the block size gets smaller is referred to as '**exponential decay**.' Exponential decay is seen often in nature. One example of this is measuring the half-life of radioactive elements in the environment. Certain substances like Uranium and Carbon-14 will radioactively decay in this pattern. Since this pattern is predictable, scientists can use it to date objects from the past.

Spring Scale

Bill and Maria are getting ready for another adventure in space that will take them to asteroids, moons, planets, and around stars.

Maria sees Bill packing a spring scale.   Maria thinks that it won't work on parts of their trip. Bill thinks it will.

Can you determine who is correct and why?

1. The spring scale will still work but it won't work at the same proportions that it does on Earth!

2. Any type of scale operates by measuring two different factors of an object. The first factor is how much mass the object contains. This is the most obvious factor to think about because with more mass you have more weight. However, what would a scale say if that mass were in deep space? The scale would not read any weight because the second factor, gravity, is not pulling that mass in any direction. Gravity is a relative force that depends on how close or far away you are from another object. On earth it seems like all gravity is the same because we are all about the same distance from the center of the earth!

3. The universal formula for gravity is shown below:

$$F = G\frac{m_1 \times m_2}{r^2}$$

This formula shows us that the force of gravity (F) is equal to the mass of the two objects (m_1 and m_2) divided by the distance between the two objects squared. So does your weight change when you are at different places on earth just like Bill's weight would change at different points in space? If you weigh 68kgs (150lbs) at sea level then the force of gravity pulling your body towards the center of the earth is 668.4N. If you were to hike to the top of Mt. Everest (8848m) then the force of gravity on you would be 666.5N, a difference of 1.9 N while still on Earth! Chances are you would not feel this difference but it shows you that your force of gravity and weight change every time you change your mass or distance from the center of the Earth!

Marble Madness #1

Maria had a large jar filled with evenly mixed red and green marbles. After an earthquake shook the house, Maria noticed that the red marbles were now all at the bottom of the jar.

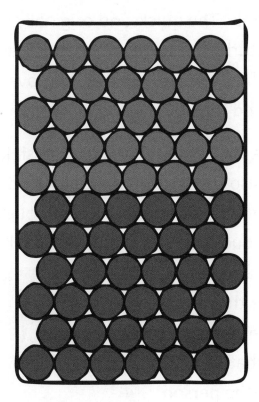

Can you determine why this occurred?

1. The red marbles were denser than the green marbles!

2. Just because an object is the same size as another object, it does not mean that it is not heavier. Think about a beach ball and a bowling ball. The beach ball is bigger than the bowling ball but definitely not heavier! When two objects of the same size and different densities are agitated the denser object will always move to the bottom of the container.

3. Quicksand operates on a similar property to these marbles. When a person walks over quicksand, the quicksand below their feet is less dense than the person's foot and therefore the foot starts to sink. The quicksand originally looks solid but once a foot disturbs the surface, the quicksand quickly becomes liquefied. The sand particles move away from the person's foot and the liquid water remains.

Density may be the reason that a person sinks into quicksand but density is also the reason that quicksand will never kill that person. Once the person has sunk far enough down in quicksand, the buoyant force of his body will keep him floating. As soon as the amount of sand and water that a person displaces equals the weight of that person's body, the person will stop sinking. Then the person will need to increase his surface area enough to break free!

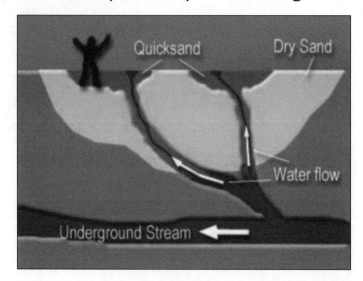

Marble Madness #2

Maria had another jar filled with evenly mixed large and small marbles. She knew that these marbles were all equal in density, so the larger marbles were heavier than the smaller ones.

Can you determine how these marbles
would react during a good shaking?

1. The smaller marbles would go to the bottom of the jar while the bigger marbles would move towards the top!

2. Since these marbles are all the same density there must be another factor in play that would cause them to separate. This factor is the size of marbles. The size of the marbles causes different amounts of extra space in between each of the marbles that are next to each other. Since the smaller marbles can fit into less space they have a higher probability of moving downwards. This pushes the bigger marbles upwards and stratifies the jar.

3. The process of petrification operates based off of the same principle as this jar full of marbles. When an organism, like a fern, dies and falls into the ground it begins to be petrified as it decays. Each cell in the fern is too big for the normal sediments in the soil to enter but over time water brings in smaller minerals which fill the space inside the fern's now dead cells. Over thousands of years those minerals build up and create the fossil that scientists can find today. The fossil is composed of materials (minerals) that are clearly different from the larger sediments that the fossil is buried in. In order  for petrification to occur, the cells of the recently deceased organism must be immediately protected from destruction due to animals or weather.

Tension Trouble

From a balcony, Bill begins lowering a string to Maria below. When the string is halfway down, Bill ties a five pound weight to it then continues lowering it.   When the string is finally within Maria's reach, she grabs it and holds tight.   Bill begins pulling on the string as Maria holds on tight.   The string's breaking point is rated at ten pounds.   Assuming that Maria will NOT be hit by the falling weight, can you determine where the string will break?

Will the string break between the weight and
Bill or between the weight and Maria?

1. The string will break between the weight and Bill!

2. In this situation the tension force on the string can be separated into two different parts: above the weight and below the weight. Below the weight the string is only experiencing the force of Maria pulling on the string. Above the weight the string is feeling the force from Maria's pull as well. However, the string above is also feeling the downward pull of the weight. These combined forces on the top string will cause it to feel more force and therefore break first!

3. This situation is relatively simple compared to the situations that physicists and engineers have to deal with in real world projects. In order to help them organize complex situations scientists use a tool called a **'force diagram.'** A force diagram is a convenient way to lay out all the forces acting on an object. The two force diagrams for the two sections of string are shown below:

Top of the String: Bottom of the String:

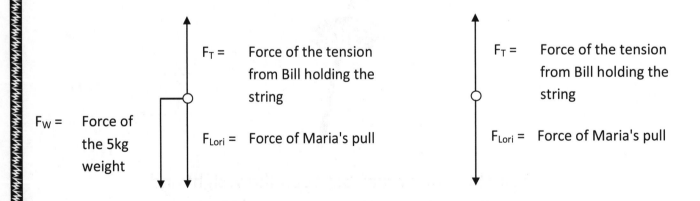

F_T = Force of the tension from Bill holding the string

F_{Lori} = Force of Maria's pull

F_W = Force of the 5kg weight

F_T = Force of the tension from Bill holding the string

F_{Lori} = Force of Maria's pull

Tension Trouble #2

Once again, Bill begins lowering a string from a balcony to Maria below. Everything is done in exactly the same way as the previous experiment.   But, just as Bill is about to start pulling, Maria yells up to him that this time the string will break on her side.

Bill doesn't believe her... but sure enough, the string breaks on the lower part!

Can you determine what happened?

1. The string breaks lower because Maria pulled the string very quickly!

2. If we could zoom in closely enough we would see that the force of tension in the string is actually just the force of atoms and molecules bonding to each other. There are trillions of these bonds in the string which hold up the 5kg weight and take the force of Maria's pull. Since Maria is pulling at the bottom of the string, the atoms at the bottom of the string are the first to experience the force of the pull. When these atoms are pulled down they pull on the atoms above them. This process would normally continue all the way up the string to Bill. However, if Maria pulls hard enough, as in this situation, the bonds of the atoms will break before they have a chance to transmit the force up the string to Bill!

3. One question that physicists are currently trying to answer is: how fast does force propagate through matter? On Earth the distances between the sides of an object are so relatively small that scientists can usually assume that a force is moving instantaneously through an object. However, what if there were a light-year long two-by-four beam in space and it was pushed on one side. Would the other side immediately move? The answer is 'no' because it would take some amount of time for all of the atoms in the beam to bounce into each other and transmit the force across space!

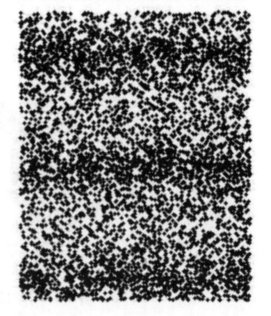

Three pulls on the string would result in three tension waves of atoms

Block Baffler

Bill constructed the following structure out of wooden blocks.

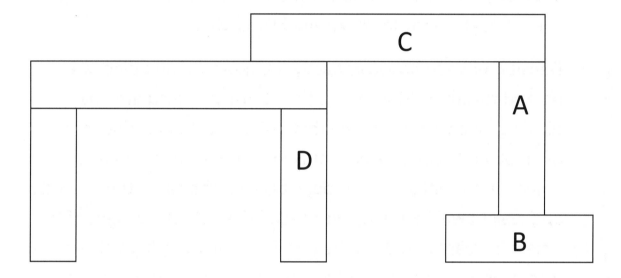

He then told Maria that he was going to remove blocks A and B.   Maria told Bill that block C would surely fall.   Bill said it would stay, and to Maria's surprise... it did!   Block C was not nailed, glued, or attached to the other blocks in any way.

Can you explain this apparent magic trick?

1. Block C is made of different types of wood so the center of gravity of the block is over block D!

2. The left side of block C is made out of Mahogany, a very dense type of wood. Mahogany is so dense that it actually sinks when put into water! The right side is made out of Balsa wood which is a lighter, less dense material. This creates a center of gravity that is skewed towards the left side of the block. If the center of gravity is over a solid object then the block will stay upright, even though it may look extremely unstable!

3. Density refers to how compactly the atoms of an object are packed together. This means that density is calculated by dividing an object's mass by that object's volume. The density of an object is commonly confused with the weight of an object. If the volume of an object is kept the same then density and weight will correlate with each other. As the weight of the object increases, so does the density. However, if an object increases in weight and volume then the density can actually remain the same. So more weight does not always mean an increased density.

Usually states of matter are less dense from solid to liquid to gas. A wax candle is an ideal example of this pattern. The solid wax is the densest, the melted wax is slightly less dense than the solid wax, and the gas emitted from the candle is much less dense than the other two states. Water is an important counter example to this pattern. Although gaseous water is much less dense than liquid water, ice (solid water) is actually less dense than liquid water causing the ice to float.

Weight Change #1

Maria was studying accurate measurements of the earth and discovered that the earth is a little flattened at the geographic North Pole and bulges a little near the equator.

Can you conclude where she would weigh more -
the North Pole, the South Pole or the equator?

1. Maria would weight more on the North (or South) Pole!

2. The Earth looks like it is shaped like a sphere. However, if you measure the Earth's height and width you will find that it is actually a more oval ellipsoid! The diameter, or distance all the way through, of the equator is 12,756.32km but the distance from the North Pole to the South Pole is only 12,715.43km. That means that when you're standing at sea level at the North Pole you are actually about 25km closer to the center of the earth than if you were standing at sea level at the equator. Being closer to the Earth's center of gravity will cause you to weigh slightly more!

3. In a gravitationally and physically perfect world the Earth would be a perfect ellipsoid. However, in reality there are multiple factors that affect the strength of Earth's magnetic field in different places and therefore the amount of gravity that people feel holding them down on Earth. The first major factor that affects our gravitational field is the different land masses that we have on Earth. More

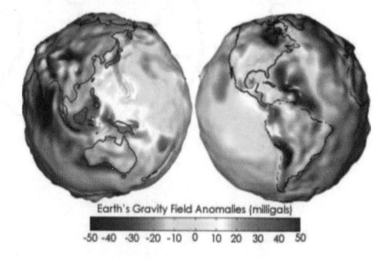

Earth's Gravity Field Anomalies (milligals)

mountains mean more mass and more gravity. Another affect is the different densities of rock in the Earth's interior. Higher densities of rock will cause a greater amount of gravity.

Weight Change #2

Bill placed a scale on a chair, then sat down on the scale. Because his legs were carrying some of his weight, the scale registered less than his full weight.

Suddenly, the scale registered only five pounds. Then it wildly changed to over twice his normal weight! The scale continued to fluctuate until Bill finally got up and walked away. Bill had kept his feet on the floor at all times.

Can you determine what was wrong with his scale?

1. The scale was working properly but Bill was using the scale on a roller coaster!

2. We feel our weight because the force of gravity is constantly pulling us towards the center of the earth. However, our weight can feel different if we are accelerating. On the roller coaster Bill first falls down a big slope. This causes him to feel barely any force pulling him down at all. But when he reaches the bottom of the drop he starts to feel much heavier than normal due to the rapid change in direction from down to up!

3. Any vehicle offers a good example of these different forces which we call 'G Forces'. When you are traveling at a constant velocity in the same direction you will not feel any force except for gravity pulling you down. However, if you turn the wheel then the car turns but your body wants to continue at the same velocity in the same direction. If you accelerate your car then you will feel a force pushing you back into your seat because the care is now moving faster than your body.

   When travelling in a car, these g forces aren't too important unless there is an accident. However, in fighter jets these forces are extreme. A human being can withstand about 5 g which is equivalent to 5 times the force of normal gravity. If the g forces get any higher than that then the heart physically can't pump blood with enough pressure to get it to the body and soon the brain shuts down from lack of Oxygen. The military actually uses special 'g suits' which help to force blood from the extremities back up into the brain. Using these suits and special training air force pilots have learned how to withstand up to 9 g!

Weight Change #3

Bill and Maria were thinking about the earth and gravity. They were wondering if they would weigh more, less, or the same if they were in a shaft, 5 kilometers below the surface of the earth.

Can you determine what their calculations would show?

1. Bill and Maria would weigh <u>more</u> than at the surface because they are closer to the center of the Earth!

2. Gravity is caused by every atom in the universe pulling on every other atom in the universe. The denser a material is, the more atoms it has in it, and therefore the more gravity it has. The surface, or crust, of the Earth is actually not very dense when compared to the mostly iron inner and outer core. This means that even though there would be five kilometers of crust pulling Bill and Maria upwards, their increased proximity to the extremely dense core of the Earth would pull them even harder downwards causing more weight!

3. Newton's Law of Gravity states that gravity is proportionate to the amount of mass of two objects divided by the radius between the two objects squared. This means that a decrease in distance between two objects has much more of an effect on gravity than an increase in mass. So, even though Bill and Maria go deeper into the Earth, they are not decreasing their gravity. However, after travelling down in the Earth far enough there will be a point where the mass of the earth above them is gravitationally equal to the mass of the earth below them and the gravity of Bill and Maria will start to decrease. If they go far enough, to the **center** of the Earth, then there will effectively be no gravity at all because gravity in all directions will cancel itself out!

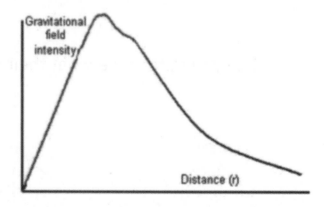

Magnifying Glass

Bill was using a magnifying glass to examine some shells and animals on the beach. After a while, Bill got thirsty and fumbled through his bag to find his bottle of water. While doing this, he dropped his magnifying class on the rocks where it shattered.

Can you determine how Bill could still get a closer look at the shells?

1. Bill could use his bottle of water to increase magnification!

2. Bill's clear bottle of water will magnify the shells because it is made out of a material other than air. The light that bounces off the shell to Bill's eye will travel at a certain speed through the air and then slow down and change direction slightly as it hits the water. This happens because the water is much denser than the air. So when light hits the different surfaces of the water bottle it will change direction to converge at one point, therefore creating a magnifying glass!

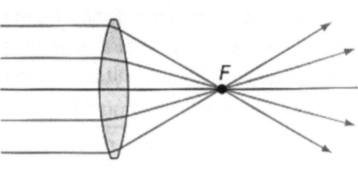

Converging lens

3. The concept of light changing directions as it moves through different media is called '**refraction**.' There are many ways to change the refraction of light in order to create a stronger or weaker magnifying glass. One way would be to change the density of the media that the light is travelling through. Glass is a much denser media than water and therefore makes an efficient magnifier.

Another way to change the magnification would be to alter the entry angle of the light. If a surface acts as a converging lens then the light will always be converging on one area, the focus. To move the focus closer or further away from the object, simply change the entry angle of the light to be bigger or smaller.

Toy Boat

Bill and Maria have a toy boat floating in a bath tub.   The tub has water level markings on the side. They place a steel ball in the boat and notice that the water level rises. Then they take the ball out of the boat and place the ball directly in the water.

Can you determine if the water level is the same or
different when the ball is placed on the boat
or directly in the water?

1. The water level will be higher when the steel ball is in the boat!

2. The water level rises due to the displacement of water. Displacement is when water is pushed by an object. Since water is liquid it wants to fill the container that it is in. This means that water will push out in the direction with the least resistance. When an object is placed in the water, the direction with the least resistance is the air above the water. Therefore the water level rises as the object is put in.

   If the steel ball is placed in the water then it will make the water level rise the equivalent of the volume of the steel ball. However, if the steel ball is placed in the boat then the water level will rise the equivalent of how much water is equal in weight of the steel ball.

3. The difference in water level is due to the volume of water being displaced. As long as the steel ball is denser than the water, the ball can only displace the same amount of liquid as its volume when it is placed directly into the water. The weight of the ball does not matter in this case. However, when the steel ball is put in the boat the ball pushes away the exact weight of water that is equal to the weight of the steel ball. This is due to the air that will also be below the water level in the boat. The weight of the air below the water's surface and steel ball will be the exact same average weight as the water that is displaced. This will be a higher amount of water than just submersing the steel ball alone because the steel ball weighs so much more than the water that it displaces.

Water Transfer

Maria had two glasses side by side on a table. One was full of water - the other was empty. Maria pulled out a non-bendable straw and told Bill that she could transfer some of the water to the empty glass... with OUT touching the glass or putting the straw in her mouth.

Bill was stumped.

Can you determine how Maria was able
to make good on her promise?

1. When the straw is partially submerged in the water Maria seals the top of it with her finger. Then she lifts the straw out of the water and puts the water into the empty glass!

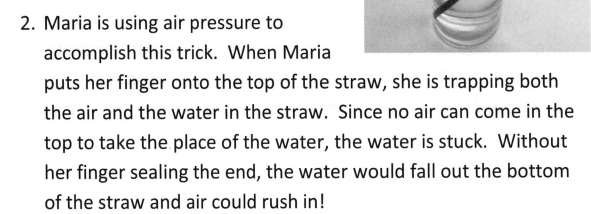

2. Maria is using air pressure to accomplish this trick. When Maria puts her finger onto the top of the straw, she is trapping both the air and the water in the straw. Since no air can come in the top to take the place of the water, the water is stuck. Without her finger sealing the end, the water would fall out the bottom of the straw and air could rush in!

3. When the straw is in the liquid without being sealed there is an equal amount of gravity and air pressure spread out across the surface of the water. When the finger seals the straw, it creates a vacuum in the top of the straw. Since the air pressure in the vacuum is much less than the air pressure still pushing down on the rest of the water in the glass, when the straw is pulled out of the glass, the water in the straw will stay in the straw while the rest of the water in the glass is pushed down by air pressure and fills the empty space where the straw used to be.

When the finger is removed from the straw then the air in the straw is again effected by air pressure and therefore releases the water into the empty glass.

Water vs. Metal

Bill and Maria placed a solid piece of metal and a solid piece of water (frozen!) in a special oven. They slowly began warming the oven up.

After some time, when the oven was hot enough, the metal melted and spread. Strangely, the water had not yet begun to melt.

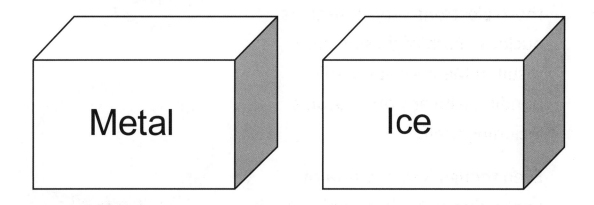

Can you determine how this could be?

1. The metal melted first because the metal is Mercury!

2. Mercury has a melting point of about -39°C (-38°F). Compare this to water which has a melting point of 0°C (32°F). Mercury's melting point is extremely low for a liquid because it is an extremely stable element. Mercury's electron structure resists bonds to other atoms. This means that Mercury behaves much like the noble gases which also have very low melting temperatures!

3. A substance either melts or stays a solid due to the strength of the bonds between its molecules. When molecules disassociate from each other the substance will lose its structure and melt amorphously.

   Mercury has a full valence electron shell ($6s^2$) and keeps these electrons very close to its nucleus. Both of these factors result in Mercury not readily bonding with any other atoms, including itself.

   Even though Mercury is fairly unreactive in its elemental state, it can cause some very serious reactions if it enters the body. Mercury reacts with Selenium in the body which is an element that is needed to make many enzymes function properly in your brain. When these enzymes do not function properly epinephrine cannot be degraded and the body goes into a constant state of adrenaline rush.

Gravity Defying Ball

Maria looks at Bill and notices that Bill is holding a string. At the end of the string is a small ball.   She watches as the ball slowly begins to move away from Bill.   Bill keeps his hand and arm steady as the ball continues to rise until the string is nearly perpendicular to his body. There is no wind on this day and the ball that is in the air is not magnetic.

Can you determine how the ball on the string is doing this?

1. Bill is on a merry-go-round that is increasing in speed!

2. The ball is being elevated by centrifugal force. If Bill were on a merry-go-round with a ball that was <u>not</u> attached to a string then the ball would go flying straight out from the merry-go-round. This shows us that the direction of the centrifugal force is straight out from the center! As the speed of the merry-go-round increases, so does the force away from the center of the merry-go-round. Eventually this force is stronger than gravity and the ball levitates in the air!

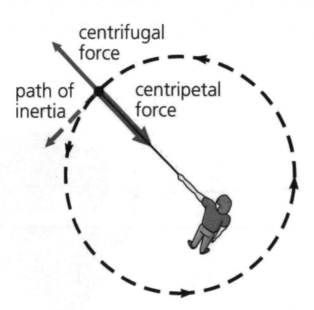

3. The formula for centrifugal force is: $a = \dfrac{v^2}{r}$

   The 'a' is the acceleration of the ball, 'v' is the velocity, and 'r' is the radius from the center of the merry-go-round to the ball. This means that the bigger the merry-go-round is, the less centrifugal force there will be. Ice skaters use the same principle when they spin starting with their hands away from their bodies moving slowly and then gradually bring their hands inwards closer to the body to get more spinning speed!

Golf Guess #1

Bill and Maria placed a normal, dimpled golf ball on a tee. A machine swung a club and hit the ball.

They did this again with a golf ball that had no 'dimples'. It was a completely smooth ball. The same machine hit the ball with the exact same force.

Can you determine which ball went further?

1. The dimpled normal golf ball went further because the dimples cause the ball to lift higher in the air!

2. When a golfer hits a golf ball, there is a certain amount of initial force put into the ball (the swing) which propels the ball in an upward and forward direction. This initial force is slowly canceled out by air molecules hitting the ball and pushing it backwards and down. Eventually all of the forward force on the ball is gone and the ball falls to the ground.

   However, if the ball has dimples then the air molecules will get 'stuck' in those dimples as the ball flies. This means that the air around the ball will cling to the surface of the ball and will create much less drag than the smooth ball.

3. The layer of air around the golf ball is called the 'turbulent layer' because that is where the air is making contact with the sides of the ball. The longer the turbulent layer sticks to the ball, the less eddies of air will form behind the ball. The eddies that form behind the ball create drag and shorten the balls trajectory through the air. Since dimples delay the air from separating, they also delay the eddies from forming and create less drag force!

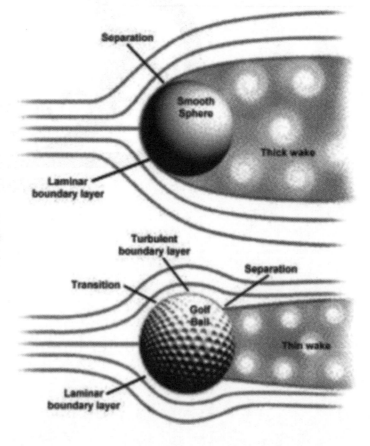

Golf Guess #2

Bill and Maria decide to do another experiment with just the dimpled golf ball. They have their machine swing a club and hit the dimpled golf ball in normal conditions.

Then they take their machine into a vacuum where there is no air (and therefore no air resistance!). Then they have the machine hit the same dimpled golf ball with the same amount of force inside the vacuum.

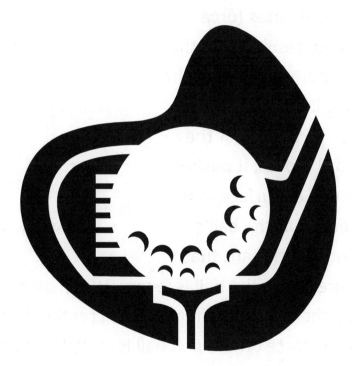

Can you determine which ball would travel further?

1. The dimpled golf ball will travel further in the normal air than in the vacuum!

2. Air resistance pushes back on a travelling golf ball in the opposite direction in which the ball was hit. This would seem to mean that the ball in vacuum would travel further than a ball in normal conditions because with no air, there is no force pushing backwards! However, when a golf ball is hit it rotates with a backspin that is about fifty revolutions per second. When air hits this backspin, an upward force is created and the ball will actually travel further up than a ball in a vacuum! This allows the ball to travel further in normal conditions.

3. The lift of a ball with a backspin is caused by **Magnus force**. Magnus force results from the top and bottom surfaces of the ball moving in opposite directions. As the top of the spinning ball moves, it pushes the air hitting its surface in the same direction as the ball's movement. This means that the pressure above the ball is lower. When air hits the bottom of the ball, the opposite effect occurs creating higher pressure. The resulting force on the golf ball is straight upwards relative to the direction of motion and creates lift. Magnus force has an effect in most ball sports including baseball and soccer. Balls thrown or kicked with a side spin will move horizontally creating non-linear paths.

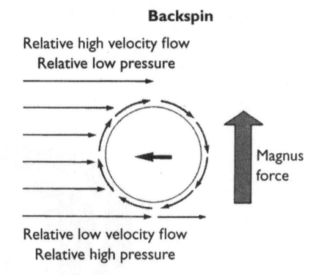

Backspin

Relative high velocity flow
Relative low pressure

Magnus force

Relative low velocity flow
Relative high pressure

Throw Up

While playing catch one day, Maria and Bill decided to do an experiment. They accurately timed a ball being thrown straight up and the time it took the ball to fall back down.

Can you determine if the ball had a longer flight up, a longer flight down, or if it took the same amount of time to go up as it did to come down?

1. The ball takes longer to come back down to the ground!

2. The answer to this problem can be found by looking at the motion of the ball. If the forces on the ball are both acting against the movement of the ball (when the ball is going upwards) then the ball will have less time in the air. If some of those forces are working with movement of the ball and some are against then the ball will have more time in the air (when the ball is falling back downwards).

Ball travelling upwards:

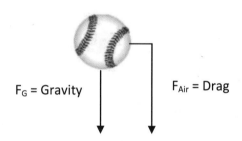

F_G = Gravity F_{Air} = Drag

3. There are two important forces involved in this problem: drag from air resistance and the force of gravity. Whether the ball is travelling upwards or downwards, the force of gravity does not change. The mass of the earth and the ball are constant therefore gravity is constant.

Ball travelling downwards:

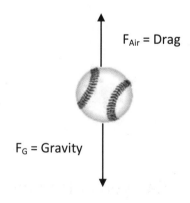

F_{Air} = Drag

F_G = Gravity

The drag of air resistance is constant as well (for a given speed). However, air resistance always resists the movement of an object. So when the ball is travelling upwards both air resistance and gravity are pulling downwards, shortening the time it takes for the ball to reach its highest point. When the ball is travelling downwards gravity is still pulling down but air resistance is now pulling upwards, the opposite direction of the ball's motion. These opposing forces cause the ball to stay in the air slightly longer as it falls back downwards!

Comet Conundrum

Bill and Maria were analyzing some data on asteroids and comets. Their studies showed that there were many asteroids and comets that could have impacted Earth. Fortunately, something was, and IS, in place that protects Earth from these asteroids and comets.

Can you determine what it is that is in place and how long it/they have been there?

1. The other planets are protecting the Earth from comets and asteroids!

2. When our previous sun exploded about 4.5 billion years ago, it created a giant cloud of rock and debris. These rocks  were attracted to each other by gravity and slowly started to condense into bigger and bigger clumps. Most of the matter strewn about the solar system condensed back into the object that we call our sun today. All of the other matter still left floating around condensed into the planets!

3. Scientists call the accretion of matter into planets the 'nebular hypothesis' and the planets of our solar system are still absorbing debris from that explosion 4.5 billion years ago in the form of asteroids and comets colliding with our atmospheres! Scientists estimate that about 100 tons of meteor material is absorbed by the Earth from space each day. The other planets of the solar systems probably have similar proportions of absorbed material. Any material 'caught' by another planet is less material that could potentially hit the Earth.

   The asteroid belt in our solar system is another good example of the Nebular Hypothesis. The asteroid belt is still an area of scattered debris because one object never achieved the mass necessary to attract all the other objects together to form a planet.

Rat Riddle

Bill and Maria set up an experiment. They put a small rat inside a large sealed container. Inside the container was a scale that the rat would occasionally stand on.

They did not give the rat any food or water for five hours. After five hours, the scale showed that the rat had lost a small amount of weight. The rat did NOT go to the bathroom during this time.

Can you determine how the weight of the rat changed even though the weight of the container did not?

1. The rat lost weight through the CO_2 (carbon dioxide) leaving his lungs into the air!

2. Everybody knows that humans (and animals) gain weight by eating food. However, many people have not thought about how that weight is lost again. Humans take in sugars and fats from food to use for energy. In these

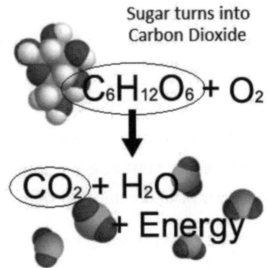

Sugar turns into Carbon Dioxide

molecules the energy is stored in long chains of carbon-carbon bonds. By the process of metabolism, these carbon-carbon chains are broken and energy is released. Once the energy is released there are many individual carbons left over. Humans use oxygen to bind to these carbons and then exhale them as CO_2 therefore losing the weight of the sugar or fats that they previously took in!

3. Breaking down sugars to obtain energy for the cell is a very complicated process called cellular respiration. Cellular respiration consists of three parts. First, glycolysis splits a normal six carbon sugar like glucose in half to create two molecules called pyruvate. This pyruvate then enters a mitochondrion and goes through the citric acid cycle which releases CO_2 as a byproduct. Finally, the energy-rich electrons that were created in the citric acid cycle are used by the electron transport chain in order to create energy. This whole process creates around 36 ATP molecules which transport the energy around the cell.

Hard of Hearing

Maria has good hearing. Yet, when Bill sneaks up behind her and yells, "Hi, Maria!" she does not flinch or turn around.   Other than a small amount of ear wax there was nothing else in or around Maria's ears.

Can you determine why Maria did not hear him?

1. Maria was at a concert and couldn't hear Bill because of all the other sounds in the room!

2. Sound travels through the air in waves. These waves have properties like how high the wave is (the amplitude) and how long the wave is (the wavelength). Changing these properties will change how the sound of the wave is perceived by Maria. So if Bill came up behind Maria and shouted with a voice that was a much lower amplitude than the sounds of the concert then Maria would not be able to hear him!

3. Changing the amplitude will change the volume of the sound but to change the pitch of the sound, the frequency must change. Frequency is the amount of wavelengths that are passing by a point in any given second. If the frequency is high then the wavelength gets shorter and you hear a higher pitch.

When you change the pitch of your voice what you are really doing is constricting or loosening your vocal cords to create different frequencies of sound waves. The vocal folds move in the throat and rapidly change the pressure of the surrounding air. This change in pressure of the air causes air to be pushed out in pressure waves which our ears can detect as sound.

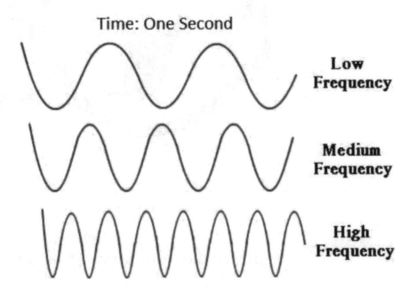

Time: One Second

Low Frequency

Medium Frequency

High Frequency

Balloon Buddies

Both Bill and Maria each brought a round balloon to a staff meeting. They placed their balloons on the table. They quickly noticed that the balloons began rolling towards each other.

Can you determine what was causing this behavior?

1. The balloons were attracted to each other due to their static electric charges!

2. Static electricity is caused by two objects being rubbed together. When the objects rub against each other, negatively charged electrons are transferred between the two. One object ends up with more electrons while the other object ends up with less. This causes one object to have lots of negative charge while the other has lots of positive charge. Positive and negative charges attract to each other so the objects will move closer together! If the objects were to touch again then the electrons could be transferred back and the charge would be lost.

3. Static electricity is 'static' because it can only build up on a non-conductive surface called an insulator. If the surface were conductive then the electrons would travel away from the surface instead of staying put.

   One way to move electrons, even on an insulated surface, is to create a spark. A spark happens when there are a large amount of electrons on a surface and, usually, another object nearby. The electrons 'jump' away from surface because they are attracted to the objects' positive protons. As the first few electrons jump away from the surface, they heat the air around them which allows other electrons to follow more easily. This process cascades until there is a spark big enough for the naked eye to see!

Coal Question #1

Bill and Maria each had a one inch cube piece of coal. The coal was the same consistency and density. Both Bill and Maria lit their cube on the same corner with the same type of match.   Yet, Maria's burned more quickly than Bill's.

Can you determine what might have caused the discrepancy?

1. Maria's coal was burning in a location that was windy while Bill's location was not!

2. Fire of any type requires three ingredients: fuel, oxygen, and a high enough temperature to start burning. If you made a campfire in the woods you would use wood as your fuel. The oxygen for your campfire would come from the air. Oxygen makes up about 21% of the air around you and cannot be increased or decreased. The only way to get more oxygen to a fire is to blow more air at the fire! This is why blowing at the base of a dying or struggling campfire can get the fire roaring again.

3. If that is the case, then why doesn't a candle's flame grow bigger when you try to blow it out? Aren't you adding more oxygen to the flame in this case as well? It is true that when you blow out a candle you are adding more oxygen to the flame. However, you are also lowering the temperature of both the wick and the wax as well. If you blow hard enough then the temperature of the wick drops below its burning point and the fire will go out.

Coal Question #2

Maria and Bill each have nine cubes of coal. All the cubes are identical. Again, they light their coal in exactly the same way. This time the air flow is exactly the same. Yet, once again, Maria's pile burns faster.

Can you determine the why?

1. Maria lined up her coal in a straight line while Bill stacked his coal in the shape of a cube!

2. By stacking her coal in a straight line Maria increased her surface area to volume ratio. This means that even though the volume of Maria and Bill's coal was exactly the same (9 cubes), Maria's coal had more of its surface touching the outside environment. More of the coal's surface touching the environment meant that there was more space for oxygen and flame to burn the coal. This made Maria's coal burn more quickly than Bill's coal!

3. The idea of surface area to volume ratio is essential to living organisms. For example, the goal of your intestines is to digest as much food as possible as quickly as possible. That is why, instead of being one long tube, your intestines are actually lined with tiny finger like protrusions called 'villi'. These villi greatly increase the surface area of the lining of the intestines which allows the intestines to absorb much more food at once.

   Another example of this concept concerns the amount of heat that an organism can dissipate out of its body. The bigger an organism gets, the smaller its surface area to volume ratio gets, and the amount of heat that it can dissipate decreases. This means that larger organisms must move slower in order to not create too much heat in their bodies. Some organisms have adapted to overcome this limitation by evolving mechanisms to get rid of heat more efficiently. Desert hares live in hot climates and have adapted by increasing their surface area to volume ratio by making their ears very large and flat. This helps dissipate heat from their body into the environment!

Snake Stumper

Two sisters were out in the desert. They stumbled upon a group of rattlesnakes and were both bitten. One sister died and one sister lived. The same amount of venom entered both girls and the venom was of the same potency.

This time Maria was stumped but Bill had an explanation.

Can you determine what Bill's analysis was?

1. One sister had built up an immunity to snake venom while the other had not!

2. Snake venom is made up of mainly neurotoxins - which attack nerves, and cytotoxins - which attack cell membranes. A small dose of either of these toxins is usually not fatal for humans because we have a small amount of proteins in our blood that can bind to the toxins and disable them. If a person, like one of the sisters, gets a small dose of toxin then her body will respond by creating more good proteins. If the dose of toxin is increased more and more then the body will continue to create more and more good proteins until eventually the sister is immune to snake bites! The other sister who is not immune does not have enough good proteins to combat the venom and therefore will die.

3. Many organisms that are the prey of venomous snakes have evolved an immunity to snake venom from the snakes in their particular environment. For example, the  Mongoose can survive a bite from a cobra which would be fatal to a human. Over generations of time, the Mongooses who had more good proteins in their blood survived and passed on their genes more often than Mongooses without those good proteins and therefore evolved an immunity to venom!

Wind Wonder

Bill and Maria were at the same location, same address, same building at exactly the same time.   The both had anemometers -- a wind speed indicator. They both opened a window on the same side of the building and at the same time took a wind speed reading.

Strangely, Bill's reading was ten mph faster than Maria's.

Can you determine why there was a difference?

1. Bill was on the 80th floor of the building and Maria was down on the 1st floor!

2. In most places on Earth, wind speed increases as you travel upwards through the atmosphere. This is due to the obstacles that wind faces closer to the surface. At ground level the air is being pushed in many directions when it hits objects in its way. This causes the lowest level of the atmosphere, the 'planetary boundary level,' to be mixed and disrupted regularly. The height of the planetary boundary level where the mixing of air occurs can vary depending on the surface structures or even time of the day!

3. Wind turbines are electrical generators that harness the force of the wind in Earth's atmosphere. Since wind speeds are higher at higher elevations, many new wind turbines are being designed to be taller and larger. The relationship between the height of the wind turbine and the amount of power generated is exponential. The amount of power generated increases quickly with only a small gain in elevation. This is shown in the diagram below:

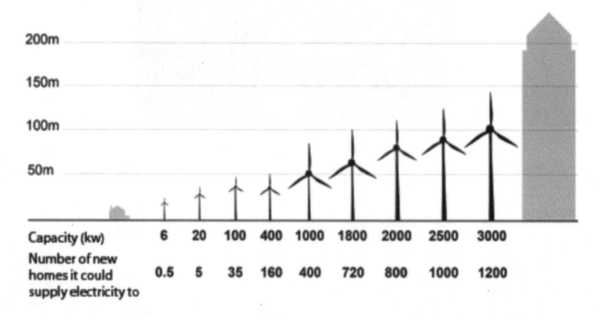

Carbon Question

Bill and Maria both picked up a three foot long - one inch diameter pure carbon rod.

Bill watched as Maria snapped her rod into two separate pieces. Try as he might, when Bill tried to break his rod, he could not. Both rods contained nothing but carbon and had identical shapes.

Can you determine why Bill was having such a hard time breaking his Carbon rod?

1. Maria's rod was graphite -- like in the lead of a pencil - while Bill's rod was made out of diamond!

2. Both graphite and diamond are made out of the same material: pure Carbon. However, the molecular bonds are different in each substance. A carbon atom can make four covalent bonds with the molecules around it. In graphite, only three of these bonds are used while the other electron floats around conducting electricity. The Carbon atoms in diamonds use all four of their bonding points to attach to the Carbons around them, making diamonds much stronger and more stable than graphite!

3. The key difference in the strength between these two substances comes from the two crystalline structures that are created due to the substance's different amounts of covalent bonds. Graphite's three covalent bonds cause a flat sheet structure which consists of many sheets stacked on top of one another. The sheets are made up of interconnected triangles of Carbon. These triangles are very strong but the sheets they make up are only connected by very weak Van Der Waals bonds.

Diamond's four covalent bonds create a tetrahedral structure. Since every bonding point is filled, this crystal structure is very strong.

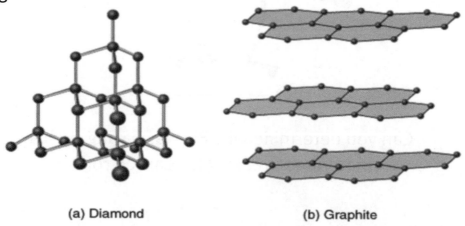

(a) Diamond   (b) Graphite

Carrot Conundrum

Bill and Maria were eating carrots. The carrots were grown for the exact same amount of time with the exact same amount of water and sunlight. But, when they were tested for nutrient content, the carrots that Bill ate had less vitamin A and other nutrients than Maria's carrot.

Can you explain why there was a difference?

1. Bill's carrot grew in soil with fewer minerals than Maria's soil!

2. The mass of a plant consists of 99.9% of these materials: **carbon dioxide** from the air, **water** from the soil, and **energy** from the sun. The other 0.1% of the mass of the plant comes from nutrients. There are six main nutrients that are needed for plant growth and many others nutrients that may be needed in extremely small quantities. Scientists call these nutrients '**micronutrients**' because they are only needed in very tiny amounts. However, without these micronutrients plants could not survive!

3. Nutrients are divided into three categories: **primary macronutrients**, **secondary macronutrients**, and **micronutrients**. Each different nutrient is used for a different purpose inside the plant. For example, phosphorus is used in the production of ATP which is the energy currency of all cells. Potassium helps to create a pump which opens and closes a plant's stomata. Stomata help to regulate how much water a plant loses through its leaves. Nitrogen is needed to build proteins which are the building blocks of cells.

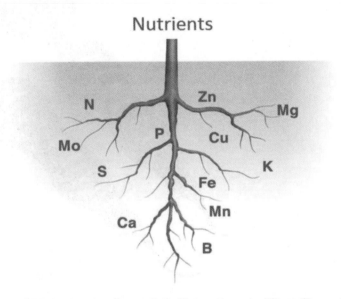

Nutrients

Macro Elements	Secondary Elements	Micro Elements
N - Nitrogen	**Ca** - Calcium	**Fe** - Iron
P - Phosphorous	**Mg** - Magnesium	**B** - Boron
K - Potassium	**S** - Sulphur	**Zn** - Zinc
		Cu - Copper
		Mn - Manganese
		Mo - Molybdenum

Coldness Conundrum

Maria came into Bill's office one day and asked him to roll up his sleeves. After blindfolding him, she placed a small cube of steel in one hand and a small cube of wood in his other hand.

Bill immediately said that the metal piece was colder than the wood.

Yet, Maria verified with a thermometer that both the steel and the wood were exactly the same temperature.

Can you explain Bill's confusion?

1. The steel cube is a better conductor of heat than the wood cube and therefore transfers heat out of the body more efficiently!

2. Heat is really just the movement of particles in an object. The faster the particles move, the more heat the object has! An  object is a good conductor of heat when its molecules can easily hit or vibrate against any molecules near them. Steel is an excellent conductor which is why we use stainless steel for pots and pans. Wood acts more like an insulator, the opposite of a conductor, and doesn't conduct heat very efficiently.

3. Scientists cannot individually measure the movement and interactions of trillions of particles in an object. Instead, they have created a unit to estimate one '**particle of heat**.' This unit is called a phonon. A phonon is one particular movement of heat at a specific frequency in an object. Most objects are much warmer than absolute zero and therefore have a massive amount of phonons interacting together.

Decreasing or increasing the amount of phonons in an object is important when designing many products. A tablet manufacturer may need to find a way to increase the amount of phonons travelling through a tablet in order to dissipate heat made by the processor. A refrigerator manufacturer may need to decrease the amount of phonons in order to stop the transfer of heat from the environment into the refrigerator.

Lake Top Test

While swimming one day, Bill and Maria watched carefully as Larry attempted to walk on the water.   Amazingly, he is able to do it!   Larry was on the surface of the same lake that Bill and Maria were swimming in.

Larry was not using any manmade support or help in any way and the lake is not frozen.

Can you determine how Larry was able to do this?

1. Larry was a water strider who uses surface tension to walk on top of the water!

2. Surface tension occurs at the surface of the water because there are more bonds between water molecules that are at the surface. Larry is putting less force down on the surface of the water with his feet than the bonds between the water molecules can handle. If Larry put all of his weight on one leg then he would fall right through the surface of the water! He must keep his weight distributed widely over the water in order to stay afloat.

3. Each water molecule can make four hydrogen bonds with other water molecules around it. These bonds are constantly breaking and reforming. A submerged water molecule will make bonds all around itself, essentially cancelling out any net force on the water molecule. However, at the surface of the water there are no bonds in the direction of the air. This creates a net force inwards and causes the 'tension' to occur on the surface. This inward force also resists gravity, creating droplets of water. Without this force the water would spread out in a more flattened shape.

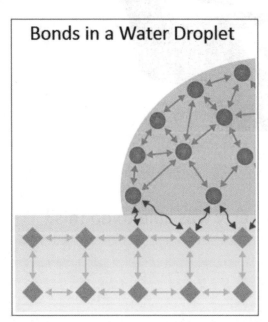

Bonds in a Water Droplet

Globe Probe

Bill and Maria were looking at an average sized table-globe.   Bill took out a ruler and placed it vertically on the globe. He asked Maria how many inches tall the atmosphere would be if it were displayed on the globe.

Maria smiled, because she knew.

Can you come close to Maria's answer?

1. The atmosphere on the globe would be less than one millimeter thick, about the thickness of the shiny varnish on the surface of the globe!

2. At a height of 100 km above the Earth, the atmosphere may seem endless but it is actually fairly small when compared to the radius of the Earth. The lowest level of the atmosphere, where all of Earth's weather occurs, is called the **Troposphere** and only extends 17 km above Earth's surface. The radius of the Earth is 6,371 km. This means that the atmosphere makes up about 1.5% of Earth's total radius and the Troposphere makes up less than .3%!

3. Scientists debate what is considered the atmosphere and what is considered space. As the distance from Earth increases the number of particles decreases. Eventually, in the Exosphere, the amount of particles is almost imperceptible.

The planet Earth has an atmosphere only because it is in the sweet spot of the solar system. If Earth were closer to the sun, like Mercury, the atmosphere would be blown into space by the intense solar radiation. If our planet were further from the sun, like Saturn, then we would have an extremely dense atmosphere because none of the gases created on the surface would be blown away. Saturn is also a very large planet and therefore has more gravity to attract its atmosphere to itself. Mercury is small and cannot hold onto any gases it creates.

Sandwich Stumper

Bill and Maria read an article about two brothers. The brothers had shared a sandwich. One brother died - one lived. It was determined that the sandwich was the cause. Both halves of the sandwich were exactly alike in every way. Bill was stumped but, Maria knew immediately what the issue was.

Can you determine Maria's conclusion?

1. The brother who died had a severe peanut allergy while the surviving brother did not!

2. Allergies are caused by an overreaction of the immune system to common environmental molecules. The immune system sees a molecule that it doesn't recognize, like peanuts or the pollen seen below, and attacks. A severe enough allergy can cause the body to enter Anaphylaxis Shock. This means that the body will swell, break out in rashes, and have very low blood pressure. A person can die from Anaphylaxis Shock due to a lack of blood to the heart or inflammation of the throat! All of these symptoms are caused by the immune system of the body and not the allergen itself!

3. When the antibodies of the immune system - called IgE - detect an allergen, they trigger a cascade of responses to deal with the problem. One of the most powerful molecules released by immune 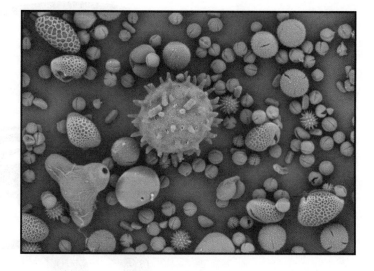 cells is Histamine. Histamine triggers the inflammatory response by allowing capillaries to become more porous. This lets more white blood cells enter areas of the body where they would normally be restricted. These white blood cells attack the allergens and cause swelling to occur. Blood pressure will also drop because more blood is in the periphery of the circulatory system.

Building Blunder

Two buildings were constructed in the same city. They were identical in every way -- same size, same shape, same materials used, etc.

Yet, when an earthquake hit, one building collapsed - the other one remained standing.

Each building was hit with the same force.

Bill and Maria analyzed the situation and identified the problem.

Can you determine what conclusion Bill and Maria came to?

1. One building was built on top of sand while the other building was on solid bedrock!

2. The land that buildings are constructed on must be evaluated for sturdiness before any building project can begin. There are multiple factors that builders consider when choosing a proper building site. The first is the risk of soil liquefaction. When an earthquake hits, the shakes can shake loose water and smaller soil particles which will be pushed up to the surface. This can cause buildings to sink into the ground. Another factor to look for is cracking. Fault lines and small cracks can open up in the event of a large earthquake swallowing buildings whole!

3. Most risk to buildings from earthquakes comes from the horizontal shaking. All buildings on Earth are designed to hold large amounts of vertical forces. However, in many third-world countries, buildings are not required to withstand the horizontal forces that earthquakes can create. When buildings are too rigid they tend to collapse easily, even in small earthquakes. This is why skyscrapers are usually built to bend a few feet in any direction, making them flexible in both wind and earthquakes. Other buildings are put on a movable base. This keeps the building stable even when the base is moving.

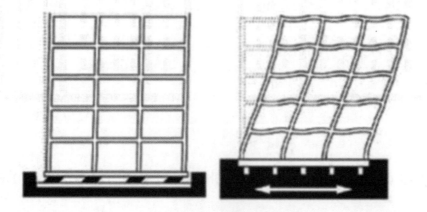

Gender Jinx

After watching the movie 'Finding Nemo,' Bill decides to study the Clownfish, a bright and colorful little fish. Bill writes in his report that he is studying a MALE fish.

Sometime later, Maria does an examination of the SAME fish. In her report she claims it is a FEMALE fish.

Can you determine the reason for their contradiction?

1. Clown Fish can change their gender when needed!

2. In a typical Clown Fish colony there is only one dominant female and one dominant male. These two fish are the only fish that can have babies in the whole colony! However, if the breeding female dies then the breeding male will turn into a female to take her place. Since the breeding male is now a female, one of the immature Clown Fish will grow quickly to take the place of the breeding male. This is the normal cycle of reproduction in a Clown Fish colony!

3. Scientists call this ability to change genders '**sequential hermaphroditism**.' A hermaphroditic change in a Clown Fish is triggered by the death of the breeding female. Normally the breeding female excretes molecules that inhibit the development of the gonads in the males around her. With the female gone, the males can discharge Gonadotropin-releasing hormone, which stimulates the development of their sexual organs.

Sweat Stumper

Bill and Maria were enjoying relaxing at a staff get-together. Suddenly, Bill began to sweat. The temperature in the room was a nice 70 degrees. No one else seemed to be uncomfortable. Bill was not wearing any extra clothing but still found himself wiping the sweat off of his forehead. He was not sick or feeling sick, nor was he exercising in any way.

Can you determine why Bill was sweating?

1. Bill was eating spicy food while Maria was not!

2. The spicy food that Bill was eating was full of a chemical called 'Capsaicin'. Capsaicin can be found in foods like chili peppers and is also used to make pepper spray. The burning effect that Capsaicin produces in Bill's mouth is caused by the chemical stimulating the pain receptors of the body. This tricks your body into thinking that it is actually being burned and triggers responses like tears, sweating, and intense pain!

3. When Capsaicin comes in contact with skin it is usually described as a burning sensation. This sensation occurs because the Capsaicin binds to an ion channel receptor that is the same receptor that is activated in intense heat or physical abrasion. When this ion channel is activated it stimulates the firing of the neuron which tells the brain that a burning sensation is happening in that area of the body.

   Capsaicin has been evolved to help limit the type of animal that eats pepper seeds. If a mammal eats the pepper seeds then the seeds are crushed by their teeth. However, if a bird eats the pepper seeds then they pass right through the digestive system! This transports the seeds to another area through the flight of the birds and deposits the seeds with a nice packet of fertilizer!

Capsaicin

Temperature Stumper

Bill and Maria were outside. Maria looked at her thermometer and the said to Bill, "Guess how warm it is?"

Bill said, "Just a minute" and pulled out his calculator. Bill said that it should be about 88° F, and he was correct!

Later, when the temperature dropped, Maria asked Bill again, and once again he was extremely accurate.

Bill did not have access to any thermometer, only his calculator.

Can you determine how Bill was able to calculate the temperature?

1. Bill listened to how fast the crickets chirped to calculate the temperature!

2. Crickets chirp for many reasons: to find a mate, to keep competitors for mates away, and even to celebrate a successful mating! Male Crickets are able to make these chirping noises by rubbing together their wings. They hold their wings high in the air while they do this in order to increase the distance their chirps are able to travel. When the temperature increases the Crickets are able to chirp more frequently. If you measure how many chirps are made in 15 seconds and then add 40 to that number you will have a good approximation of the outside temperature in Fahrenheit!

Number of Cricket Chirps in 13 Seconds

Cold: 15

Cool: 25

Warm: 37

Hot: 49

Time: Two Seconds

3. Crickets are cold blooded. This means that the temperature of their bodies is determined by the temperature of their surroundings and how fast their bodies are working. If the temperature of their environment is low, then enzymes in their bodies will not have the proper

activation energy that is needed to perform chemical reactions, which slows the body's movements and processes.

Amazing Alex

Bill and Maria were watching Alex lift some items. Suddenly, Alex attempted to lift an item three times his own weight. Amazingly, he was able to do it and carried the item some distance. Bill and Maria were stunned.   Alex was not in space and gravity was normal.

Can you explain Alex's amazing strength?

1. Alex is an ant who can lift fifty times his own weight!

2. Ants have incredible strength due to how small they are. The smaller an animal's total volume, the stronger the animal tends to be relative to bigger animals! In fact, the strongest animal scientists have found when compared to other animals is the dung  beetle. The dung beetle can drag more than 1000 times its own weight in dung! That amount of weight is equivalent to a human adult male pulling a space shuttle!

3. When measured gram for gram humans and ants actually have the same strength in their muscles. However, ants can lift fifty times their own weight due to their muscle's high cross sectional area compared to their total muscle volume. The strength of a muscle is determined by the amount of cross sectional area that the muscle has. This is the reason that human muscles grow bigger in girth and not in length. As an organism increases in size, their relative strength decreases. This means that if an ant were the size of a human it would have roughly the same amount of strength as us!

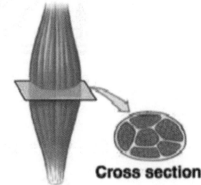

Cross section

Under Sea Level Stumper

Bill and Maria read an article about a woman who had gone down 50 meters below sea level... and died.

When they brought the body to sea level they found that there was no water in her lungs. She had not been wearing any scuba gear or anything else over her mouth. She had not been killed by any ocean creature.

Can you determine how water was kept out of her lungs?

1. She died in Death Valley which is up to 86 meters below sea level at some points!

2. Death Valley is sandwiched between the Black Mountains in the east and the Sierra Nevada Mountains in the west. The reason that Death Valley is below sea level is because it is sinking into a fault that runs along the Black Mountains. Normally, locations with lower than sea level elevations will fill up with water. However, Death Valley stays dry because of its extremely arid conditions. The less than five centimeters of rain that pour onto Death Valley each year evaporate quickly into the atmosphere leaving Death Valley very dry!

3. Up until about 250 million years ago, Death Valley was filled with a large inland sea. When the climate in the area started to get warmer, the sea started to evaporate away. This left large areas of salt from the salt water and limestone from sea creatures that can still be seen in the rocks today!

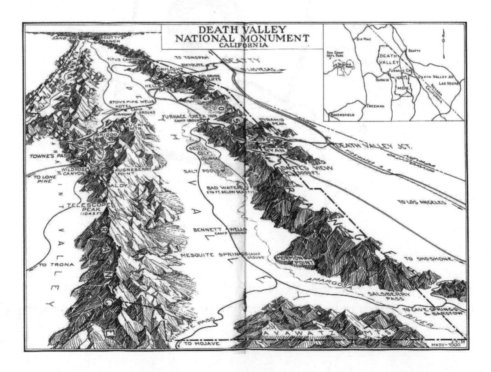

Chemical Question

Bill was watching Maria as she was preparing a meal.

"Wait," he said, "don't put that in your recipe! It has over 50 different chemicals in it!"

Maria acted as if she hadn't heard him and added it anyway.

Can you determine Maria's lack of concern and
what it was that she was adding to her meal?

1. Maria was adding an orange to her dish!

2. Everything in nature is built up of chemicals. Besides the obvious sugars and vitamins, an orange also contains many other chemicals that can be used in the body. Limonene, for

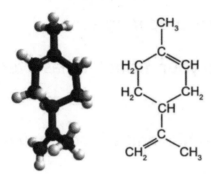

Limonene

example, is a chemical that helps food pass through the digestive system more easily. This same chemical can be isolated in a laboratory and used for other purposes like in cleaning solutions, perfume, and even 3-D printing!

3. Many chemicals found in nature have been used in creative ways by scientists. In fact, an estimated 70% of all medicines that we use today are in some way derived from plants. Plants and animals have been evolving together for over a half a billion years. This co-evolution has led to many chemical interactions between plants and animals. For instance, many berries contain Theobromine. This chemical is similar to caffeine except that it also lowers the blood pressure of any mammal that eats it. This means that mammals avoid these berries while birds are able to eat them without ill effects. Through its use of chemicals, the plants have selected which type of animal can eat their seeds. By selecting organisms that can spread their seeds further without destroying them with teeth, the plants have increased their evolutionary fitness and have a better chance of passing on their genes to the next generation. This is all due to the Theobromine that they have evolved to produce!

Bone Baffler

Bill used an x-ray machine to count all of the bones in his body. He found out that he has 206 bones! He then x-rayed Maria and found that she had the exact same amount, 206. But when they used the x-ray machine on Pablo, his count came in at over 300 bones!

Pablo is a male human and there is nothing considered abnormal about him.

Can you determine the cause of the difference?

1. Pablo is a baby!

2. When a baby is born it can have up to 350 bones in its body. However, this number is a little misleading. Babies actually have all of the bones that an adult would have but many of the bones simply have not fused together. The areas in between these fused bones are composed of cartilage. As the baby grows older this cartilage is turned into bone in a process called '**ossification**.' Calcium is added to the bones until they become hard and rigid.

Human Skull Sutures

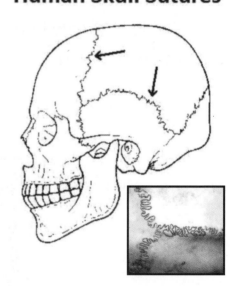

3. A human skull is an excellent example of bones that do not fuse together until later in life. An infant like Pablo would start off with five bony plates that make up their skull. As the infant grows, these plates fuse along four major sutures to create one large skull bone.

Infants need multiple plates instead of a skull in order to pass through a woman's birth canal. During childbirth, a baby's skull is compressed into a different shape than normal. This allows the baby to have a larger head later in life to accommodate the human adaptation of increased intelligence. The large heads that are needed to support this intelligence are an evolutionary trade-off. The bigger our heads evolve to be, the more intelligence they can achieve but they also kill more women in childbirth. Before modern medicine, death by childbirth was a leading cause of death for women all over the world.

Funny Froggy

Bill had mistakenly left his pet frog outside. Since it was winter, he found his frog the next morning frozen. He was sad that his frog had died. He took the frog inside. Surprisingly, a few minutes later, the frog began to move again!

Can you explain how this frog survived?

1. Bill's frog was a wood frog and can withstand being frozen for several weeks!

2. Wood frogs have several mechanisms that help them survive the freezing process. First, they sequester a majority of their body's water to a few key organs. These organs, like the large intestine and lymph nodes, are less likely to be damaged by ice formations. Second, the frogs create ice nucleation points, places where ice can form. By letting ice form in intercellular areas, the frog can keep its precious cell contents safe. Lastly, wood frogs produce massive amounts of glucose (sugar) and urea (a chemical in urine) to lower the freezing temperature of their bodies. All of these measures combined allows wood frogs to survive freezing up to 65% of their bodies multiple times each winter!

3. Some organisms take survival even further. Tardigrades are a single phylum of animals that can survive even in the harsh, emptiness of space. When their environment changes, they

can initiate states of cryobiosis and anhydrobiosis where their bodies become resistant to freezing and water loss until their conditions improve!

Maddening Mold

Bill and Maria noticed a small amount of mold growing on their two pieces of bread. The slices were from the same loaf and were alike in every way.

When they examined the slices three days later, Maria's slice was completely covered with mold but Bill's slice barely had more mold on it than three days previous.

Can you determine why there was a difference?

1. Bill kept his slice of bread in the refrigerator!

2. Mold, like all living organisms, has certain optimal conditions for growth. The optimal conditions for mold are warm and moist. Bill's slice of bread was in the refrigerator and was therefore colder than Maria's slice. The chemical reactions occurring in the mold occurred more slowly in the mold on Bill's slice of bread. This slowed the growth of the mold. Refrigerators work to keep food fresh because they put pathogens like mold and bacteria out of their optimal conditions!

3. The human immune system also works to keep pathogens out of their optimal conditions. Pathogenic bacteria that infect humans have evolved to have the same optimal conditions as the human body. They like a temperature of 37°C (98.6°F). The immune system responds to a bacterial infection by raising the body's temperature, inducing a fever. Raising the temperature of the body is not good for the infected person but the immune system is betting that the increased  temperature will be even worse for the invading bacteria. The bacteria struggle to carry out normal enzymatic reactions and other chemical functions or their metabolism. This leads to weakened bacteria which can be more easily destroyed by the human immune cells.

Maddening Mold #2

Bill and Maria once again have two slices of bread containing a small spot of mold.

When they checked the bread three days later, both slices were nearly covered with mold. Yet, the mold on Maria's slice had a red tint to it. The mold on Bill's slice had a blue tint to it.

Can you determine what was making the difference?

1. The two slices of bread had different kinds of mold on them!

2. Molds are one of the most diverse categories of organisms on Earth. Scientists estimate that there are over 400,000 different types! Different molds can have different colors, shapes, textures, and smells. However, molds do have a few characteristics in common as well. All molds are multicellular fungi, which means that they work to break down organic material for food. They also reproduce by creating tiny spores which can float through the air. This is why food can grow mold even if nothing but air has ever touched its surface!

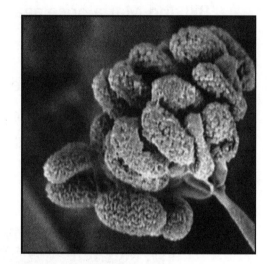

3. Molds also have many beneficial functions in the environment. Humans use molds to make many of our favorite foods like Roquefort and Bleu cheeses. Molds are also used to create antibiotics. Penicillin, the world's first antibiotic, is created by the mold *Penicillium*.

Perhaps the most interesting use of mold is to prevent the overpopulation of one species. If a species gets to prevalent in one area, commonly in the rain forest, then it is more likely to get infected by a mold pathogen and drop back down in numbers. An example of an ant is shown in the adjacent picture.

Digestion Question

Maria was upset that her dog had swallowed one of her rings. When she told Bill about the incident he replied, "Don't worry, the ring NEVER entered the dog's body."

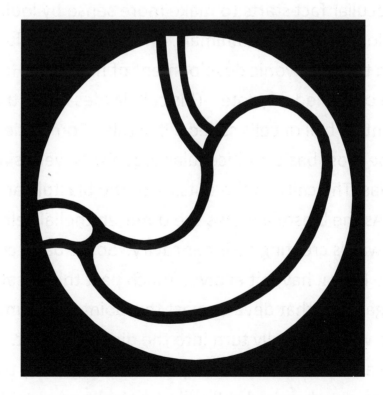

Assuming they are talking about the same ring,

can you explain Bill's response?

1. The inside of the dog's digestive tract is actually the outside of the dog's body!

2. All mammals have four different types of tissue. Muscular and nervous tissues make up the muscles and nerves of the body. Connective tissue makes up the fat, bone, blood, and more. The last tissue is called epithelial tissue. Epithelial tissue lines the outside of the body. However, the digestive tract is also made up of epithelial tissue meaning that it too is technically the outside of the body!

3. This peculiar fact starts to make more sense by looking at the evolutionary history of animals. Much of this evolution can be seen in the embryonic development of humans. Humans start as one cell called a zygote. This cell divides a few times until it turns into a ball of cells called a blastula. Some scientists think that the most basic multicellular organisms were similar to blastulas. The cells on the outside of the blastula are epithelial cells. As the blastula grows the outer epithelial cells start to fold inwards creating an inner cavity. Some basic creatures, like the hydra, haven't evolved much past this basal structure. For organisms that develop past this point, like humans, this in folding will eventually turn into the digestive tract. Nutrients can be absorbed into the body from the digestive tract but anything not absorbed will still technically be outside the body and will be excreted out of the anus.

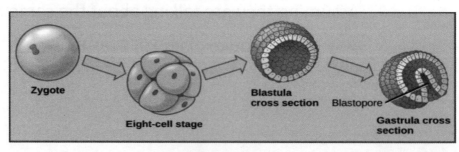

Magic Water

Bill and Maria read about a man who had a cut on his finger. The man put his finger in a river. When he lifted his hand, the cut was completely gone.

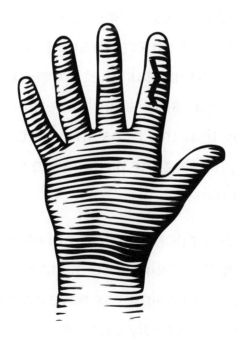

Can you determine what RIVER the man put his finger into?

1. The man's cut was gone... and so was his finger! He dipped his hand into the Amazon River and it was eaten off by piranhas!

2. Most people think of piranhas as attack fish that swarm humans and bite them to death. In reality, piranhas rarely bite humans. When they do bite, it is usually out of desperation due to a lack of food in the dry season when the river water level is low. Piranhas are omnivores, and although they will bite other fish and mammals from time to time, like humans, their main diet consists of plant material and insects. Piranhas also school together in the dry season. Scientists believe that this behavior is actually defensive rather than aggressive, helping the piranhas to avoid predators like caiman and dolphins!

3. Since piranhas are opportunistic omnivores they have very highly tuned senses. Their sense of smell is similar to that of a shark. Both sharks and piranhas have an olfactory membrane where water is continuously passing over as they swim. They use this membrane only for detecting chemicals in the water, including proteins contained in blood.

The piranha's sense of smell is very sensitive. However, contrary to popular belief, piranhas will not swarm to a drop of blood as soon as it hits the water. The blood needs time to spread out in the water by diffusion.

Black Hole Debate

Maria said she was going up to the observatory to LOOK at some black holes. Bill said he was going up to the observatory to LISTEN to some black holes.

Can you determine which statement was correct?

1. Bill is correct, black holes cannot be seen but they can be 'heard!'

2. Black holes are collapsed stars that have so much gravity that even light can't escape and gets pulled inside. If light is pulled in then light is also not escaping  and Maria will have no light hitting her eyes with which to see the black hole. However, black holes 'spit out' high energy particles from their poles that cause pressure waves to spread through the gas surrounding the black hole. These waves can be 'heard' because they release x-rays at the same intervals as the pressure waves. It turns out that these waves are an extremely low note, 57 octaves below middle C!

3. Sound cannot propagate through space. Since there are no air molecules in space, there is no medium for the sound pressure waves to travel through. However, scientists use the x-ray observatory called Chandra to detect the frequency of x-rays emanating from distant black holes and other phenomena. Then they can synthesize the frequency of the x-rays to the frequency of sound waves in order to create the 'sounds' of the black hole. These 'sounds' are well below that range of human hearing in their basic form but can be adjusted further in order to fall into our audible range.

Sun Screen Stumper

Bill and Maria were sent to a science conference in Mexico. The first day there they were scheduled to go study reptile life in a nearby desert.

Maria put on some SPF 15 sunscreen protection. Bill put on none but said, "Now we have the same sun protection."

Only considering bare skin, can you explain how Bill was able to make this claim?

1. Bill has darker skin than Maria!

2. Darker skin contains more of the chemical '**Melanin**' than lighter skin. This difference equates to about a 15 SPF advantage when in the sun. Melanin absorbs Ultraviolet (UV) rays from the sun before they can damage the DNA of the epidermis. Humans have the ability to produce more Melanin when exposed to the sun, which causes a tan! However, if the UV rays penetrate deep enough into the skin then a burn will result.

3. Every human has roughly the same number of melanocytes in their skin, the cells in which melanin is produced. The differences in skin color result from different types of melanin being produced. **Pheomelanin** has a lighter color while **eumelain** takes on a darker pigmentation. The amount of melanin in each cell can vary greatly between ethnicities as well. Even within one human, the forearms and scalp each have about twice the amount of melanin seen in other parts of the body, like the abdomen. All of these differences in melanin levels are the result of human adaptations to avoid the damaging ultraviolet rays of the sun.

Melanocytes and melanin in various skin types

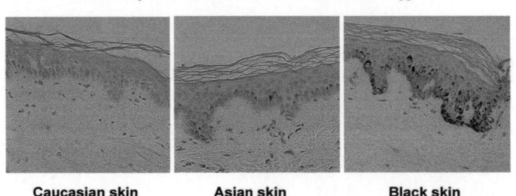

Caucasian skin   Asian skin   Black skin

Source: The National Library of Medicine

Radio Wave Riddle

Bill and Maria were working on a new problem. They were calculating how much total energy humans had ever collected from the large radio wave telescopes (over 100 in the world). They took that calculation and used it to determine how long that energy would keep a 60 watt light bulb burning.

Can you determine a close estimate to the length of time that the light bulb would stay on?

1. The light bulb would stay on for less than a second!

2. The amount of energy contained in a wave of light is determined by the light's frequency and wavelength. If the wavelength gets shorter than the frequency of the waves will rise and the amount of energy in the wave will increase. Therefore waves at the longer end of the electromagnetic spectrum, like radio waves, will have much less energy than shorter waves, like visible light. Astronomers use these longer radio waves to detect objects like pulsars and newly formed nebulas which would be harder to locate using visible light.

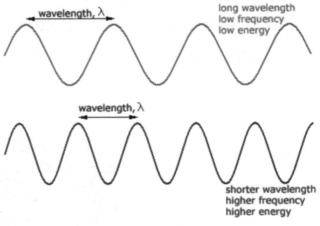

Electromagnetic Waves

3. There are two equations that help scientists determine how much energy is in a wave. The first shows that frequency multiplied by wavelength always equals the speed of light. If you know the frequency of a wave then you must know the wavelength as well, and vice versa. The second equation shows that the energy of the light is equal to Plank's constant multiplied by the speed of light and divided by the light's wavelength.

Equation #1: $f * \lambda = c$
Equation #2: $E = {}^{hc}/_{\lambda}$

Variables:
f = frequency in Hertz (Hz = ${}^{cycles}/_{sec}$)
λ = wavelength in meters (m)
c = the speed of light (299,792,458 ${}^{m}/_{s}$)
E = energy in electron Volts (eV)
h = Plank's constant (6.626 × 10^{-34} ${}^{m}{}_{2}{}^{kg}/_{s}$)

Rocket Riddle

Maria and Bill both launched a rocket 100 meters apart.   The rockets were exactly the same in every way and the engines produced exactly the same thrust.

Yet, when they reached orbiting altitude, Maria's rocket was traveling at a higher velocity than Bill's rocket! They did not pass through any clouds or hit any objects.

Can you determine why Maria's rocket was faster than Bill's rocket?

1. Bill's rocket was fired in the direction of Earth's spin while Maria's was fired in the opposite direction.

2. Velocity is the speed of an object in a given direction. To measure speed and direction, we use the Earth as a reference point. Technically both Bill and Maria's rockets are going the same speed in relation to the Sun. However, from the perspective of a person on Earth, Maria's rocket is going much faster because her rocket went in the opposite direction of the Earth's spin. This means that Maria's rocket would appear to have the speed gained from her engines (just like Bill's rocket) AND the speed of the Earth's spin (about 465 m/sec).

3. This situation illustrates the idea of relativity. Speed can only be measured relative to another object. If there were no Earth to act as a reference point then there would be no speeds of objects as we measure them today.

   Einstein's theory of relativity is operating off of a similar concept. The general theory of relativity states that space and time are relative quantities, not absolute quantities. This means that time and space can change depending on your velocity. If Maria were to travel at almost the speed of light around the galaxy and then come back to Earth, she would barely have aged at all while Bill would be decades older than when she left! Their time frames are relative to their velocities in relation to each other!

Dehydration Contemplation

The average adult needs about 3 liters of water a day to stay healthy. That means that one person would need 1095 liters of water to survive for a year. But, when Bill and Maria were asked to list all the items needed for a 6 month trip to Mars and the 6 month trip back to Earth, they only listed 10 liters of water!

Can you determine why Bill and Maria listed such
a small amount of water for their trip?

1. Bill and Maria will recycle the water in their urine!

2. 1095 liters of water weighs about 2464 lbs. NASA estimates that each pound of material launched into space costs about $10,000. That means that it would cost $24,640,000 just to launch a year's worth of water into space for one person! Average urine contains about 95% water. If that water could be recycled out of the urine, then the amount of water needed for a yearlong space flight could be reduced to just a few liters. This would save millions of dollars!

3. Humans will die without water within three days. However, there are some organisms that can go their whole lives without drinking. Desert animals, like the Kangaroo Rat, and many insects don't require any extra water at all. They obtain all of the water that they need from their food sources in two ways. First, these organisms absorb any liquid that is in their prey such as blood or saliva. Second, they can extract water from the sugar that they eat.

In heterotrophs, organisms that don't make their own food, oxygen is used to extract energy from sugar molecules. The products of  this metabolic reaction are carbon dioxide and water. Carbon dioxide is exhaled as a waste product but the water is used inside the body. This water is referred to as 'metabolic water' because it is formed by the breakdown of food. Organisms that do not lose water from their bodies easily can retain this metabolic water and use it as a major source of hydration.

Sunlight Stumper

Bill and Maria were asked how long it takes sunlight to reach the Earth from the Sun. Bill said light takes around 8 minutes to travel the 152 million kilometers (94.5 million miles) between the Earth and the Sun. Maria said it took a hundred thousand years!

Who do you think is the closest?

1. Bill and Maria are both right but the answer depends on the starting point of the light!

2. The speed of light is about 300,000 km/sec (186,000 mi/sec). Since Bill knows that the distance from Earth to the sun's surface is 152 million km, he can easily calculate that light takes about 8 minutes and 20 seconds to travel to Earth from the sun. This means that if the sun magically disappeared then you wouldn't notice for about 8 minutes!

   However, there is another way to view this problem. Maria is measuring the time light takes to reach Earth starting with when the light is created. Photons of light created in the core of the sun bounce around for anywhere from 10,000 to 200,000 years before reaching the surface!

3. Energy emitted from the sun is originally created in the sun's core by a process called nuclear fusion. Nuclear fusion uses the tremendous force of gravity at the center of the sun to force two Hydrogen atoms together. These two Hydrogen atoms fuse together and form one Helium atom while also releasing a tiny bit of their mass (about .7%) as energy. This energy is first emitted in the form of gamma rays which are absorbed into the sun's plasma and then released again almost instantaneously. This process occurs until the photons randomly arrive at the surface of the sun and shine into space.

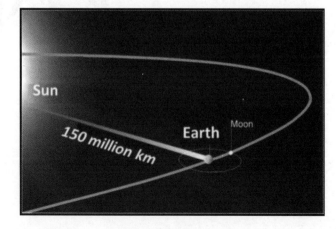

Sailing Stumper

Bill and Maria were asked by their boss to design a new spaceship. When they presented a spaceship with a large sail on it, the boss laughed. "There is no AIR in space! This design won't work!" he said.

Can you determine what Bill and Maria were thinking when they made this design?

1. Bill and Maria were using the solar wind to power their spaceship!

2. Unlike the Earth, space is completely devoid of air molecules. However, there are still particles permeating almost every part of our  solar system. These particles are photons of light from the sun! When these energy filled particles hit an object in space they transfer a tiny bit of energy to that object as they bounce off and fly away. This energy is very tiny but over time all of these photon impacts will add up and be able to push even a very large object. This method of propulsion is possible because there is no friction in space since there is nothing to slow the spaceship down. Therefore, every photon impact will move the spaceship a little bit faster!

3. Photons propel spaceships with solar sails in two ways. First the impact of the photon pushes the spaceship. Second, the photon bounces off the sail in the opposite direction creating another force. Making solar sails that have a reflective surface will increase this second force. Since the sun is emitting photons out into space through the 'solar wind' at all times, the bigger a solar sail is, the faster it will travel through space. As long as the sun is shining, the solar sail and spaceship will continue to accelerate!

Categorized Index

Astronomy

Biology

Chemistry

General Science

Geology/Earth Science

Meteorology

Physics

For new books and games, visit us at: MissingPiecePress.com